Motion Planning

Edited by Edgar A. Martínez García

Published in London, United Kingdom

IntechOpen

Supporting open minds since 2005

Motion Planning
http://dx.doi.org/10.5772/intechopen.94620
Edited by Edgar A. Martínez García

Contributors
Zahra Elmi, Soheila Elmi, Zain Anwar Anwar Ali, Muhammad Shafiq, Eman H. Alkhammash, Edgar A. Alonso Martínez García, Santiago de J. De Jesús Favela Ortiz, Angel P. del Pobil, Emanuele Sansebastiano, Valeri Kroumov, Kenta Takaya, Hiroshi Ohta, Keishi Shibayama, Abdul Majeed, Seong Oun Hwang

First published in London, United Kingdom, 2022 by IntechOpen
IntechOpen is the global imprint of INTECHOPEN LIMITED, registered in England and Wales, registration number: 11086078, 5 Princes Gate Court, London, SW7 2QJ, United Kingdom
Printed in Croatia

British Library Cataloguing-in-Publication Data
A catalogue record for this book is available from the British Library

Additional hard and PDF copies can be obtained from orders@intechopen.com

Motion Planning
Edited by Edgar A. Martínez García
p. cm.
Print ISBN 978-1-83969-773-9
Online ISBN 978-1-83969-774-6
eBook (PDF) ISBN 978-1-83969-775-3

For EU product safety concerns:
IN TECH d.o.o., Prolaz Marije Krucifikse Kozulić 3, 51000 Rijeka, Croatia, info@intechopen.com or visit our website at intechopen.com.

Meet the editor

Dr. Martínez is a full professor at Universidad Autónoma de Ciudad Juárez since 2007; he founded and currently leads the robotics laboratory and the academic body of mechatronics at the Institute of Engineering and Technology. He received a BSc in computing engineering systems and an MSc in electronics engineering. Professor Martínez obtained his Ph.D. in robotics engineering from the University of Tsukuba, Japan in 2005. He worked as a research fellow for Nanyang Technological University, Singapore (2005–2007) and as a postdoctoral fellow at Advanced Materials Research Center, Mexico (2007–2008). His research interests are dynamics modeling and control of robots, planning and navigation, scientific computing for bio- and neuro-robotics systems, sensors and actuators, and control of mechanisms and machines.

Contents

Preface

Motion planning is an interesting and inherent area in most robotics research and development. It ranges from a local motion control of a single actuator to a complex mobile machine compounded of sophisticated capabilities, such as path generation, trajectory tracking, global autonomous navigation, wide-space coverage for inspection and exploration, and local or global pathfinding in unknown environments. This book has organized a selection of recent advances in chapters divided by three general perspectives: a) aerial path planning and tracking, b) artificial intelligence (AI)-based optimization for planning, and c) wheeled robots planning and control. The chapters present specific planning works, essentially on estimation, prediction, optimization, observation, and control in dynamic scenarios.

The first section introduces concepts and methods for unmanned aerial path planning and tracking that are critical for prospective new complex paradigms that contribute to the continued resilience of airspace coverage and applications performed by unmanned aerial vehicles.

The second section presents AI approaches, heuristic and meta-heuristic optimization solutions for planning. Heuristic methods are problem-solving algorithms that emulate human thinking. Traditional graphical maps used for path planning when combined with heuristic approaches perform faster than some deterministic solutions, providing feasible paths. Moreover, optimization algorithms inspired by biological entities show fast convergence and efficiency in estimating locomotion for dynamic robotic agents. Bioinspired methods might perhaps solve multi-objective optimization problems, multi-robot path planning, formation control, and self-organization of distributed systems.

The third section treats one of the most traditional types of robotic platforms, the wheeled robots, which rely on a wide range of navigation control techniques. Wheeled vehicles depend on path planning and tracking to reach certain autonomy when facing the environmental dynamics to which they are subjected to. Local planning performance is critical due to multiple unpredictable obstacles in order to attain safe and suitable pathways for self-driving vehicles. A robotic platform depends on its kinematic mobility constraints to considerably perform path following. Longitudinal and lateral controls are required to efficiently track complex polynomial routes. Some mechanical structure designs depend on active and passive motions (e.g., limb bar linkage), where numerous analytical solutions are dominated by algebraic and derivative methods to model dynamic local planning considering obstacles, goals, and routes, simultaneously.

Finally, I would like to acknowledge the author service manager Ms. Sara Debeuc for her editorial support to complete this book. Additionally, I would like to thank the anonymous external reviewers for providing valuable comments to improve the chapters' technical quality.

Edgar A. Martínez García
Instituto de Ingeniería y Tecnología,
Universidad Autónoma de Ciudad Juárez,
Juárez, Mexico

Section 1

Aerial Path Planning and Tracking

Chapter 1

Recent Developments in Path Planning for Unmanned Aerial Vehicles

Abdul Majeed and Seong Oun Hwang

Abstract

Unmanned aerial vehicles (UAVs) have demonstrated their effectiveness in performing diverse missions at significantly lower costs compared to the human beings. UAVs have the capabilities to reach and execute mission in those areas that are very difficult for humans to even reach such as forest, deserts, and mines. Integration of the latest technologies including reactive controls, sense and avoid, and onboard computations have strengthened their dominance further in various practical missions. Besides the innovative applications, the use of UAVs imposes several challenges, and one of those challenges is computing a low-cost path for aerial mission by avoiding obstacles as well as satisfying certain performance objectives (a.k.a path planning (PP)). To this end, this chapter provides a concise overview of various aspects concerning to PP including basics introduction of the subject matter, categorization of the PP approaches and problems, taxonomy of the essential components of the PP, performance objectives of the PP approaches, recent algorithms that have been proposed for PP in known and unknown environments, and future prospects of research in this area considering the emerging technologies. With this chapter, we aim to provide sufficient knowledge about one of the essential components of robotics technology (i.e., navigation) for researchers.

Keywords: unmanned aerial vehicle, path planning, low-cost path, algorithms, aerial missions, urban environments, time complexity, coverage path planning

1. Introduction

In recent years, unmanned aerial vehicles (UAVs) have become a powerful tool for diverse missions including polymerase chain reaction (PCR) samples transportation between hospital and laboratories [1], UAV-based healthcare system to control COVID-19 pandemic [2], infectious diseases containment and mitigation [3], traffic condition analysis in co-operation with deep learning approaches [4], and human behavior understanding via multimedia data analytics in a real-time [5], to name a few. Currently, UAVs integration with the emerging technologies such as block chain, internet of things, cloud computing, and artificial intelligence can pave the way to serve mankind effectively compared to the recent past [6]. Further, the peculiarity of UAVs in terms of performing operations in 3D (dull, dirty, and dangerous) environments, they can play a vital role in realization of the smart cities.

Furthermore, UAVs are inevitable tool during emergency planning and disaster management due to their abilities to perform missions aerially. Besides the UAVs applications and use cited above, they can be highly beneficial for military purposes including information collection and analysis, border surveillance, and transporting warfare items. The role of UAVs in agriculture from multiple perspectives have already been recognized across the globe. Recently, world's leading commerce company (i.e., Amazon) has started using UAVs for delivering their products to customers. Generally, the use of UAVs is expected to rise in many emerging sectors in the near future. We present actual and innovative use of the UAVs during the ongoing pandemic in **Figure 1**. Majority of the applications given in **Figure 1** employed multiple UAVs in order to accomplish the desired tasks.

Although UAVs are highly beneficial for mankind through their innovative applications, but there exist plenty of challenges that can hinder their use at a wider scale. For example, payload constraints and power issues can limit their carrier abilities. Similarly, decision making during flight to ensure UAVs safety by avoiding

Figure 1.
Innovative applications of the UAVs during the ongoing pandemic (adopted from [7]).

obstacles with sufficient accuracy is a non-trial task mainly due to no human-onboard control. Furthermore, communication from long distances, and co-ordination among multiple UAVs to perform complex tasks jointly are main barriers in the true realization of the UAVs technology. Besides the challenges and issues given above, many issues concerning software and hardware also exist that need rigorous developments and testing. Many solutions have been proposed to address these issues via cross disciplinary approaches. Meanwhile, extensive testing and analysis of these solutions is yet to be explored, especially in urban environments. In this chapter, we mainly focus on the 'navigation' that is one of the core challenges in the UAVs technology. The navigation quandary is classified into three cases: (i) where am I now?, (ii) where do I go?, and (iii) How do I get there?. The first two cases belong to the localization and mapping, and the third case is about path planning (PP) [8]. In this work, we cover third case comprehensively, and provide concepts and developments in this regard. We present a comprehensive overview about changing dynamics of the UAV applications in recent times,

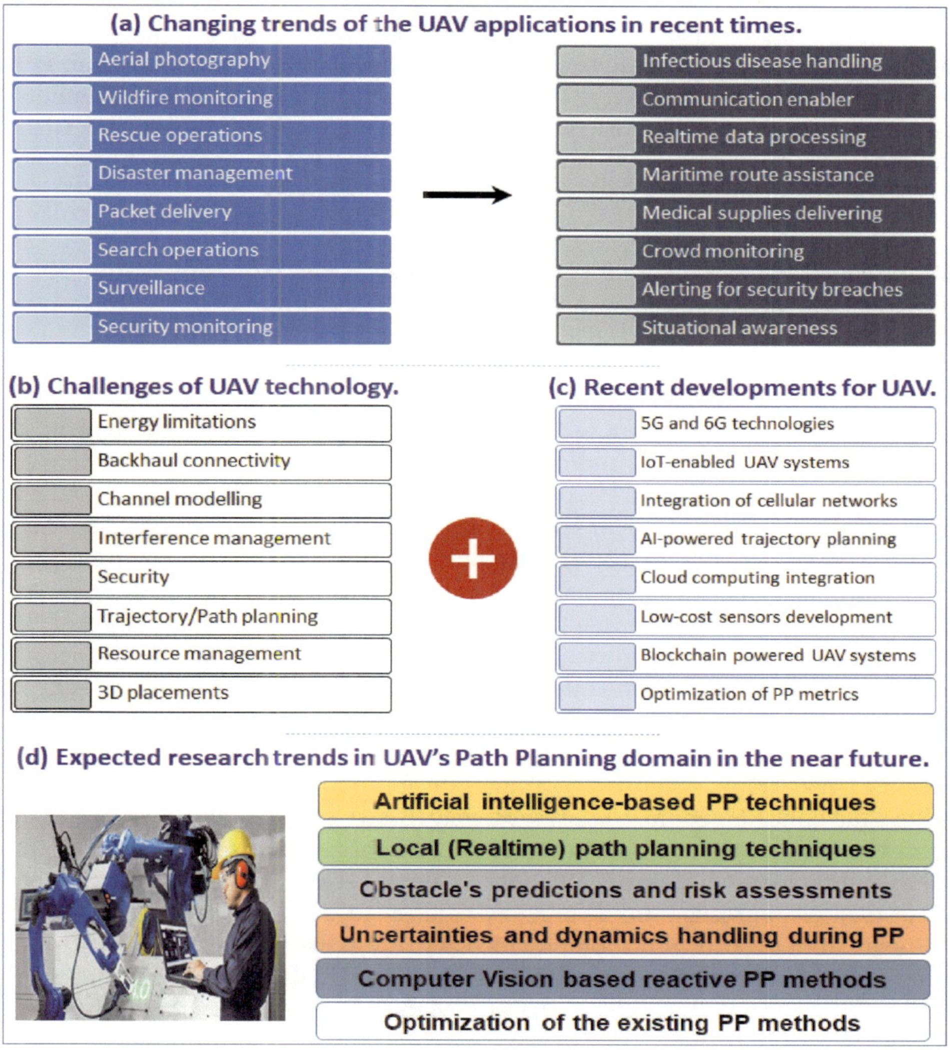

Figure 2.
Overview of changing dynamics of the UAV applications, challenges, recent developments, and future research trends in the PP area.

challenges of the UAV technology, recent developments in the UAV technology, and future research trends in the PP area in **Figure 2**. With this concise overview, we aim to aid researchers in extracting the contents enclosed in this chapter conveniently.

The rest of this chapter is structured as follows. Section 2 discusses the basic concept of the path planning, and categorizes the path planning approaches based on the information available about underlying environment, and UAV used for the aerial mission. Section 3 describes the three essential components of the PP. Section 4 critically analyzes various approaches that were proposed to lower the computing time of the PP for UAVs. The future prospects of the research in the PP area are discussed in Section 5. Finally, this chapter is concluded in Section 6.

2. Path planning and categorization of the path planning approaches

PP is to find a safe (i.e., collision-free) path between two pre-determined locations (e.g., source and destination, denoted with s and t, respectively) by optimizing certain performance objectives. The performance objectives can be energy consumption, computing time, distance, path smoothness, and turns etc. depending upon the mission type, operating environment, and UAVs' type. The most important part of the PP is to identify the environment where the pathfinding is carried out for UAVs. In this work, we categorize the PP approaches based on the type of environment's information, and UAVs strength, respectively.

2.1 Categorization of the path planning approaches based on information about environment

Generally, there are three possibilities about the availability of information regarding environment where UAVs tend to operate. The operating environment can be fully known in advance (e.g., obstacles' geometry information is known.), it can be completely unknown, and/or it can be partially known (e.g., few portions are known, and some portions are explored and modeled during the flight.). Based on the degree of information about environment, PP approaches are mostly classified into two categories, local PP (LPP) and global PP (GPP). In LPP, the environment is not known, and UAVs use sensors or other devices in order to acquire information about the underlying environment. In GPP, PP is performed in a fully known environment, meaning all information about environment is known in advance. Based on the availability of the information regarding underlying environment, GPP approaches have lower complexity compared to the LPP approaches. Recently, some PP approaches have jointly employed LPP and GPP concepts in order to find a path for UAVs [9]. In literature, GPP and LPP approaches are also classified as offline and online PP approaches, respectively. Based on the extensive review of the literature, we present a categorization of the PP approaches based on information about environment in **Figure 3**. We refer interested readers to gain more insights about the LPP approaches in the previous studies [10, 11].

Apart from the categorization provided above, environment can be classified into rural and urban environments. The tendency of UAVs applications were high in the non-urban environments in the past. Moreover, due to the significant development in control domain, UAVs are increasingly employed in the urban environment these days. For instance, in urban environments, they can be used to monitor people compliance with the social guidelines given by the respective governments in order to control the COVID-19's spread.

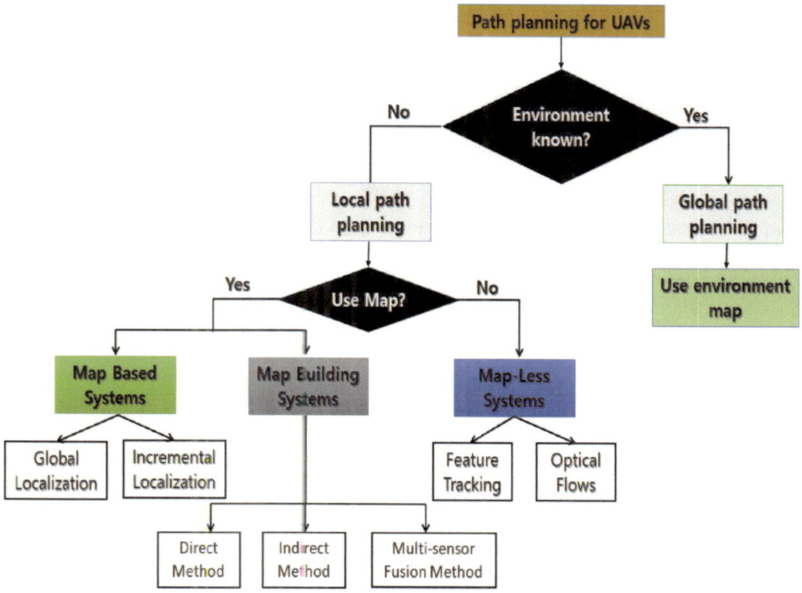

Figure 3.
Categorization of the PP approaches based on the availability of information about operating environment.

2.2 Categorization of the path planning problems

Based on the mission's type, either one or multiple UAVs can be employed. The scenarios in which only one UAV is deployed are referred as single agent PP problem. In contrast, those scenarios in which multiple UAVs are used are called multiple agent PP problems. PP for multiple agents is relatively complex since UAVs need to avoid collision with the companion UAVs, and obstacles present in an underlying operating environment. In addition, allocating target areas for coverage and optimizing throughput also remain challenging, especially while operating at lower altitudes in urban environments.

3. Essential components of the path planning for UAVs

Generally, there are three essential components of the PP: (i) modeling of the environment with geometrical shapes by utilizing the obstacles/free spaces knowledge provided by a real-environment map, (ii) task modeling with the help of graphs/trees keeping source and target locations in contact, and (iii) applying search algorithm inclusive of the heuristic function to determine a viable path.

3.1 Modeling of the environment with geometrical shapes

In the first step, a raw environment map is converted into a modeled one, in which obstacles are represented with the help of geometrical shapes. For example, poles information provided by a real environment map can be modeled with the help of cylinders in the modeled map. Similarly, buildings can be modeled with the help of rectangles or polyhedron. In some cases, UAVs do not model the whole environment map, and utilize sense and avoid (SAA) abilities to operate safely in

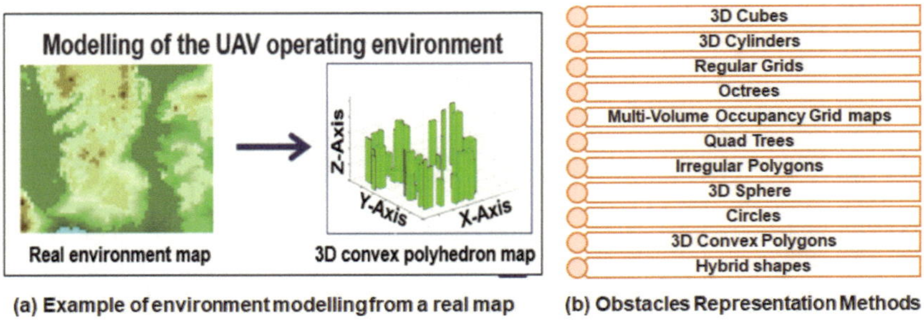

Figure 4.
Overview of environment modeling and obstacles' representation techniques.

the airspace. We present an example of environment modeling, and well-known obstacles' representation techniques used for the PP in **Figure 4**. Each obstacles representation technique has different complexity and accuracy in terms of real environment obstacles representations. In addition, each representation can be adopted considering the UAV operating environment. For example, polygons can be used to model an urban environment populated by various buildings.

3.2 Task modeling with the graphs/trees

After modeling environment with the help of geometrical shapes, the next step is task modeling (e.g., generating network of paths with a graph/tree or selecting a desired portion to be modeled). For example, road-map approach is a well-known task modeling approach for the PP, in which a graph is constructed from the starting location to destination location by capturing the connectivity of free spaces and obstacles' corners. Apart from it, cell-decomposition and potential field are promising solutions for the task modeling. We present most widely used task modeling methods in **Figure 5**.

Recently, trees-based task modeling methods have been widely used for the task modeling due to their quick convergence in the final solution. We present an overview of the task modeling with the help of tree in **Figure 6**. Furthermore, in some cases, more than one methods are jointly used to model the tasks on a provided map. In addition, some approaches use task modeling and path searching simultaneously [12].

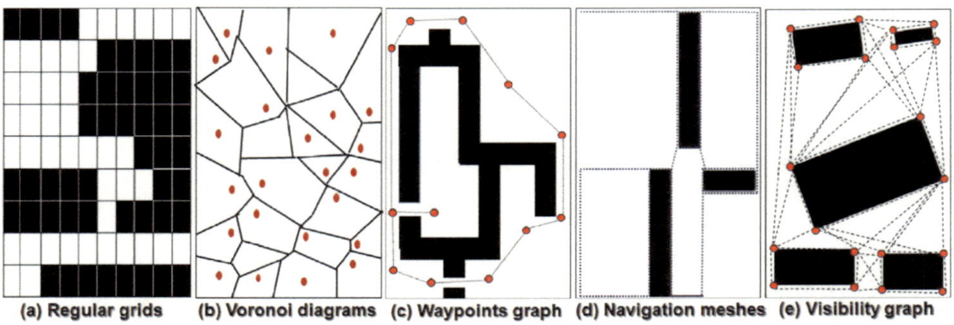

Figure 5.
Overview of the famous task modeling methods used in the PP adopted from [26].

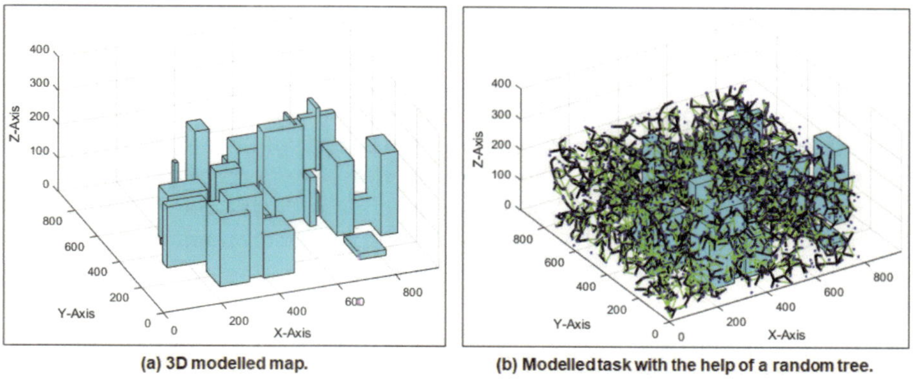

Figure 6.
Overview of task modeling with a random tree.

3.3 Applying path search algorithm to determine a viable path

In the last step, a search algorithm is employed on the graph/tree to find a viable path. During the path search, a heuristic function usually accompany the path search. For example, in the A* algorithm, the low-cost nodes are determined leveraging distance as a heuristic function. Similarly, the heuristic function can be energy consumption or smoothness depending upon the scenario. In literature, many techniques have been suggested to find reliable paths. The path search algorithms, such as differential evolution [13], firefly algorithm [14], ant colony optimization [15], genetic algorithms [16], artificial bee colony [17], particle swarm optimization [18], fuzzy logic [19], central force optimization [20], gravitational search algorithm [21], simulated annealing [22] and their advanced variants are used in the PP. Every algorithm has numerous distinguishing factors over others regarding conceptual simplicity, computational complexity, robustness, and convergence rates etc. We categorize the existing path search methods into five categories, and present representative methods of each category in **Figure 7**.

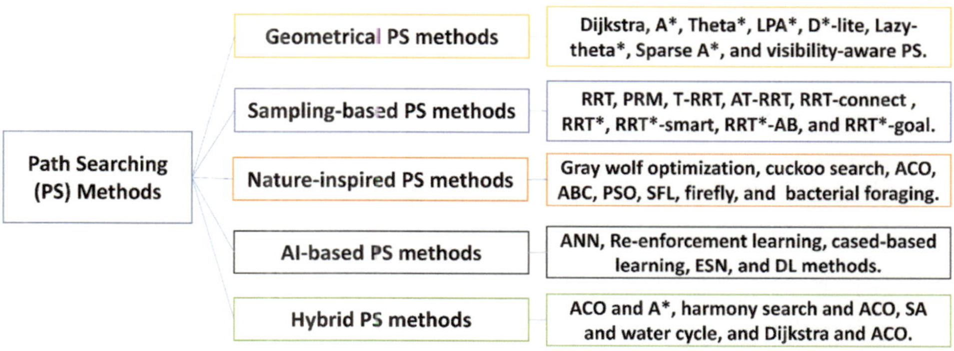

Figure 7.
Categorization of path searching methods/algorithms.

3.4 Performance objectives of the path planning approaches

Every PP approach tends to optimize one or more performance objectives (PO) while finding a viable path for UAVs. The PO can be related to hardware and software. These PO are considered in the previous three components (i.e., environment modeling, task modeling, and path searching) related to the PP. For instance, in order to lower the PS computing time, only some portion of a map can be modeled and a sparse tree/graph can be constructed/used while finding a path. Similarly, memory can be preserved by exploring some portions of a graph/tree rather than loading and exploring whole graph/tree at a same time. The selection of PO solely depend on the nature and urgency of the mission. For example, in search and rescue missions, the PO can be path computing time in order to reach the affected regions quickly. In contrast, in normal circumstances, the PO can be the path length in order to reach the target location in a most economical way by preserving UAV's resources. We describe various most commonly used PO in **Table 1**.

These PO are usually considered during PP irrespective of the environment whether it is known or unknown. Furthermore, plenty of techniques have been

PO	Concise description
Computing time	It denotes overall time required to find a path using a graph/tree.
Path length	It denotes the Euclidean distance between two locations.
Energy	It denotes amount of energy required/consumed while reaching to target from source.
Turns	It denotes number of turns (infeasible curvature) a path has in total.
Smoothness	It denotes a turns in a path with a feasible curvatures.
Memory	It denotes amount of memory used while computing a path.
Path nodes	It denotes set of nodes that a UAV follows during flight.
No. of obstacles	It denotes set of obstacles to be processed during path search.
Accuracy	It denotes accuracy of obstacles modeling or path clearance from obstacles.
Problem size	It denotes size of problem on which path is determined.
Graph size	It denotes size of graph (no. of nodes, edges) employed to find a path.
Convergence rate	It denotes how quickly a feasible solution can be obtained.
Constraints handling	It denotes the effective resolution of constrains UAV faces during mission.
Completeness	It denotes availability/non-availability of solution in a finite time.
Flexibility	It denotes efforts/time required to make a solution usable for different missions.
Path re-configuration	It denotes efforts/time required to gain the control of a lost path.
Path following	It denotes the ability to keep following a path despite disturbances.
Path safety	It denotes the ability to avoid collisions with static/dynamic obstacles.
Hyper parameter	It denotes the number and variety of parameters to find a path.
Obstacle avoidance	It denotes the ability to avoid static/dynamic obstacles with low-cost.
Generalization	It denotes the ability of a method to be applicable for different types of UAVs.
Application-speciality	It denotes the ability of a method to yield superior performance in some context.
Endurance	It denotes the ability of a UAV to fly for a long period of time with low-cost planning.

Some PO are positively co-related. For example, finding path with less turns can save energy.
Improving two negatively co-related PO (speed and time) require optimization of another PO (problem size).

Table 1.
Overview of the PO improved by the PP approaches.

proposed to improve these PO with innovative techniques or employing cross-disciplinary concepts. In addition, many PP approaches have targeted optimizing multiple objectives rather than one/two for practical UAVs application. These PO can be expressed as a functional model while finding a path P between two locations s and t. Some algorithms tend to optimize more than one POs. The overview of two PO to be optimized by a PP approach is mathematically expressed as follows.

$$\begin{cases} P = [s = p_1, p_2, p_3, \dots, p_n, p_{n+1} = t] \\ Minimize\, f_1(P) = Path\ \ Length(P) \\ Minimize\, f_2(P) = Computation\ \ Time(P) \end{cases} \tag{1}$$

4. Path planning algorithms that were proposed in the past five years

In this section, we discuss various PP algorithms that were proposed to lower the time complexity of the PP process. We selected various algorithms that were proposed in last five years (i.e., 2016–2021), and have somewhat identical concepts in terms of space restrictions and problem size reduction etc. We provide brief overview, and technically evaluation of all algorithms and highlight their deficiencies. Consequently, this analysis can pave the ways to improve PP algorithms for future UAVs' applications.

4.1 Global path planning algorithms

4.1.1 Brief overview of the selected path planning algorithms

We present brief overview of the selected algorithms in **Table 2**. These algorithms have become state-of-the-art for many practical applications of the UAVs in the urban/non-urban environments. They are famous due to their novel working mechanisms, and conceptual simplicity. In addition, they have mainly focused on the UAV applications in urban environments that is focus of research across the globe. Also, the UAVs' applications in the urban environments are likely to increase in the coming years.

Ref.	Publication year	Environment used	PO improved
Maini et al. [23]	2016	3D	Computing time and collision-free paths.
Frontera et al. [24]	2017	3D	Computing speed and solution quality.
Ahmad et al. [25]	2017	3D	Computing speed and energy-optimized paths.
Majeed et al. [26]	2018	3D	Computing speed and path quality.
Han et al. [27]	2019	3D	Feasible paths with reduced time.
Ghambari et al. [28]	2020	3D	Computing time and memory consumption.
Majeed et al. [29]	2021	3D	Computing speed and path quality.

All these approaches have used concepts related to search space reduction in order to find time-efficient paths.

Table 2.
Overview of the latest GPP approaches that were proposed to reduce the computing time of PP process.

4.1.2 Technical evaluation of the selected path planning algorithms

In this subsection, we provide concise description of the selected algorithms, and highlight their technical problems. We mainly describe the key steps of the proposed algorithms.

- Maini et al. [23] algorithm computes a low-cost path using two-steps approach. In the first step, modified version of the Dijkstra algorithm is used to find an initial path. In the second step, initial path is optimized more by considering the initial path nodes, and reverse path search.

- Frontera et al. [24] algorithm computes a low-cost path using three-steps approach. First, the proposed method reduce the search space by considering the obstacles that are on the straight axis between s and t. Later, a visibility graph is generated solely from the corners of the selected obstacles. In the last step, A* algorithm is employed to compute a shortest path incrementally.

- Ahmad et al. [25] algorithm computes a low-cost path using four-steps approach. Firstly, search space is bounded using obstacles of the straight line only. Later, the bounded space is extended to next level by using the obstacles that hit the boundary of the first bounded space. In the third step, a relatively dense visibility graph is generated from the bounded spaces. In the final step, A* algorithm is employed to find an energy-optimized path.

- Majeed et al. [26] algorithm computes a low-cost path using five-steps approach. First, the space is reduced into a half-cylinder form with path guarantees between s and t. In the second step, multi-criteria based method is employed to check the suitability of the reduced space for low-cost pathfinding. Later space is extended if needed, and sparse visibility graph is generated that ensure connectivity between s and t, and path is computed. Moreover, in some cases, path is improved by adding more nodes around the initial path's nodes.

- Han et al. [27] algorithm computes a low-cost path using three-steps approach. First, critical obstacles are identified through straight-axis between s and t. In the second step, a node set is generated around the corners of the critical obstacles only. In the last step, a feasible path is obtained by exploring nodes set. This approach is beneficial by resolving constraints related to obstacles shapes.

- Ghambari et al. [28] computes a global and local path with the help of four-steps. In the first step, search space is reduced around the straight axis. In the second step, differential evolution algorithm is applied to construct a graph. Later, A* algorithm is used to find a path from a graph constructed in the first step. In the third step, subspace is divided into small portions with alternate routes in each subspace. In the last step, a mechanism is suggested to avoid collision with the dynamic obstacles that may appear unexpectedly during the flight.

- Majeed et al. [29] recently proposed a PP method for low-cost pathfinding for UAVs based on the constrained polygonal space and a waypoint graph that is extremely sparse. In proposed approach, search space is restricted into a polygonal form, and its analysis is performed from optimality point of view with the help of six complexity parameters. Later, space can be extended to next level if needed, else a very sparse graph is generated by exploiting the visibility, far-reachability, and direction guidance concepts. The suggested approach computes time-efficient paths without degrading path quality while finding paths from urban environments.

Besides the computing time, these algorithms can indirectly optimize certain PO listed in **Table 1**. For example, Ahmad et al. [25] PP approach reduces the number of turns also in order to lower the energy consumption. Han et al. [27] PP approach can be applied to the environments with arbitrary shaped obstacles (e.g., there exist no constraint related to the obstacles' geometries). Hence, it can be applied in different settings (e.g., areas with sparse obstacles or areas with dense obstacles) of the urban environment. Similarly, Majeed et al. [29] PP approach can significantly reduce the problem size, thereby memory requirements can be magnificently lower. Ghambari et al. [28] approach can be used to re-configure paths during the flight when a UAV finds an unexpected obstacle. Hence, this approach can be used in both (i.e., local, and global) environments. Despite the utility of these approaches in many real-world applications, they often yield poor performance due to the local/global constraints. Based on the in-depth review of all studies, we identified potential problems of all approaches that may hinder their use in actual deployment. We describe technical challenges of the existing approaches in **Table 3**.

Ref.	Technical problems in the proposed approach
Maini et al. [23]	The performance cannot be ensured in each scenario due to heavy reliance on specific maps. Overheads can increase exponential with the problem size. It models the whole map thereby path exploration cost is very high.
Frontera et al. [24]	Path can collide with the nearby obstacles. In some cases, proposed approach fails to find a path even though it exists. Visibility graph can contain many needless and redundant nodes. Memory consumption is higher due to loading of whole visibility map in the memory.
Ahmad et al. [25]	Two bounded spaces are used that can increase the computing time of the PP. Visibility graph is constructed using layered approach with many redundant nodes and edges. Visibility check function is expensive since visibility in all directions and nodes is checked.
Majeed et al. [26]	Path can contain turns due to the strict boundary of the search space. Path optimization cost may increase if initial path has many nodes.
Han et al. [27]	Path quality cannot be ensured in all scenarios if obstacles' sizes are large. Path cost can increase exponentially with the point set. Both time and optimality can be impacted if diverse shape obstacles exist in a map. Since this is grid-based approach thereby memory consumption is higher.
Ghambari et al. [28]	Path computing time can rise with the distance between s and t. Recognition and avoiding obstacles in realtime can be costly. Fidelity of the proposed approach were analyzed with limited testing. Since path searching is carried out twice, thereby computing time can rise.
Majeed et al. [29]	Accurate modeling of the tiny obstacles is not possible. Excessive calculations are performed in space analysis thereby complexity can rise.

All these problems have been highlighted by existing studies or reported by the authors.

Table 3.
Overview of the technical problems in the proposed GPP approaches.

These challenges lay foundation for the future research in the UAVs area. Furthermore, they can assist researchers to devise better and practical PP approaches in order to address these technical problems. Apart from the challenges provided in **Table 3**, it is paramount to take into account the local constraints while devising PP methods that have been mostly assumed in the existing approaches.

4.2 Local path planning algorithms

Majority of the approaches discussed above are the GPP approaches, and LPP approaches have not been discussed. To cover this gap, we discuss various representative LPP approaches in **Table 4** along with the methodological specifics.

These approaches perform PP in environments that are mostly unknown, and are complex compared to the GPP approaches. These approaches enable UAVs to perform tasks in complex environments in real time leveraging low-cost sensors, and robust artificial intelligence (AI) techniques. In addition, these techniques have abilities to co-work with the emerging technologies including cloud, edge, and fog computing etc. for variety of applications. The role of UAVs was dominant during the ongoing pandemic in different countries across the globe. To this end, LPP approaches contributed significantly, and enhanced UAVs role in curbing the pandemic spread via online missions. Barnawi et al. [47] proposed an IoT-based platform for COVID-19 scanning in which UAVs were used as a main source of temperature data collection in the outdoor environments. Apart from the COVID-19 scanning, UAVs were extensively used for spraying and disinfecting multi-use facilities and contaminated places. In some countries, they were used for alerting

Ref.	UAV used	Technical aspects of the approach
Stecz et al. [30]	Multiple	Indicated sensors based LPP approach.
Wojciech et al. [31]	Single	EO/IR systems and SARs based navigation.
Siemiatkowska et al. [32]	Multiple	MILP based LPP using EO/IR camera and SARs.
Hong et al. [33]	Multiple	MILP-based multi-layered hierarchical architecture.
Hua et al. [34]	Multiple	Multi-target intelligent assignment model based LPP.
Cui et al. [35]	Single	Reinforcement learning (RL)-based LPP approach.
Maw et al. [36]	Single	Graph and learning based LPP approach.
Wei et al. [37]	Single	Improved ACO for LPP.
Zhang et al. [38]	Single	Markov decision process (MDP) based LPP approach.
Zammit et al. [39]	Multiple	LPP in the presence of uncertainties.
Wu et al. [40]	Single	Interfered fluid dynamic system (IFDS) based LPP.
Bayerlein et al. [41]	Multiple	Multi-agent reinforcement learning (MARL) approach for LPP.
Jamshidi et al. [42]	Single	LPP based on improved version of Gray Wolf Optimization.
Yan et al. [43]	Single	Sampling based LPP approach in urban environments.
Sangeetha et al. [44]	Single	Gain-based dynamic green ACO (GDGACO) LPP approach.
Sangeetha et al. [45]	Single	Fuzzy gain-based dynamic ACO (FGDACO) LPP approach.
Choi et al. [46]	Single	Improved CNN based LPP approach for UAV.

All these approaches have used the unknown environment during the PP.

Table 4.
Overview of the latest LPP approaches used for UAVs.

people to wear masks properly, and stay indoors. The true realization of these innovative application is possible through LPP approaches.

4.3 Coverage path planning: a subtopic of the path planning

Besides the LPP and GPP, another important subtopic of the PP is coverage path planning (CPP) [48]. In the CPP, a path is determined that enables UAV to cover a target area fully with the help of a device/tool mounted on it. The attached tool/ device can be a sensor, camera, speaker, and/or a spray tank depending upon the mission. We present overview of the CPP in **Figure 8**. In **Figure 8(a)**, a target area in the form of a rectangle is given that need to be covered with a UAV. In In **Figure 8(b)**, a coverage path is shown that a UAV follows in order to cover the target area.

In the CPP, most of the POs are identical with that of the PP, but path overlapping, and coverage guarantees are two additional POs. Moreover, ensuring consistent path quality with respect to shape of the target area is very challenging. Therefore, shape of the target area is considered while finding a coverage path. CPP can be performed in five steps, modeling of the operating environment, locating target area on the modeled map, decomposition of the target area into disjoint sub parts, task modeling (mainly traversal order of the sub parts) with the help of a graph, and covering each sub-part using motion pattern (e.g., back and forth, spiral, and circular etc.). In recent years, UAVs' coverage applications in the urban environments have significantly increased, and a substantial number of CPP approaches have been proposed [49].

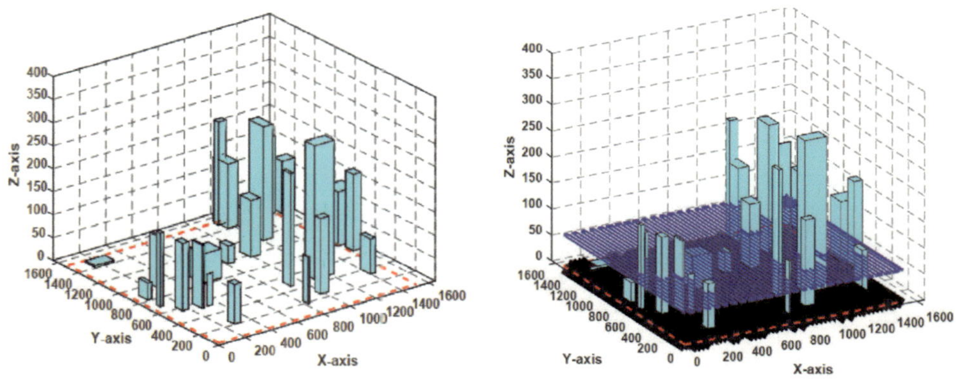

(a) Area of interest located in a 3D map. **(b) Coverage flight path (Blue lines).**

Figure 8.
Overview of coverage path planning for UAVs in a 3D urban environments.

5. Prospects of the research in the near future in the PP area

In the near future, UAVs will be regarded as an inevitable tool for various practical missions, especially in the urban environments. A substantial number of developments are underway to fully realize smart cities, smart infrastructure, and smart buildings, to name a few. Thence, the use and applications of the UAVs are expected to grow significantly in the near future. Recently, many innovative technologies such as block-chain, IoT, 5G/6G technologies, and deep/machine learning approaches have been integrated with the UAVs technology to serve mankind in effective ways [50]. For example, BloCoV6 scheme [51] is one of the wonderful applications of the UAVs in the new normal (e.g., COVID-19 era). Similarly, many such innovative applications are likely to emerge in the near future as a replacement of human beings for complex tasks. Therefore, refinements in the existing PP approaches in relation with peculiarities of the applications/tasks, and development of robust approaches leveraging cross-disciplinary (e.g., biological inspired, AI-powered, and technology-driven) concepts have become necessary. Considering the emerging applications of the UAVs, we list prospects of the research in the near future in PP area in **Figure 9**. We categorize the avenues of future research in the PP area on four grounds (e.g., UAV application specific PP approaches, optimization of the existing approaches' PO, integration of the emerging technologies and their issues handling, and developing PP approaches that can cope up with the dynamics of the UAV operating environment.).

The most important research avenues from the optimization point of view are, devising new environment restriction methods to reduce the problem sizes, devising low-cost methods for reducing the task modeling overheads (i.e., graph/tree sizes), and accelerating the PS methods that enable UAV to reach the target location safely with a significantly reduced cost. Furthermore, improving overall cost of the PP process is an important research direction to increase UAVs' applications in the urban environments. Optimization of multiple objectives rather than single/two is handy in order to preserve UAV's resources during aerial missions. From applications point of view, low-cost methods that can improve certain POs and can satisfy the applications features at the same time are needed. To this end, identifying each application's features/requirements and embedding them into the PP process can enhance the UAVs use in the coming year significantly. Therefore, applications-oriented PP methods will be embraced more in the near future considering the UAVs potential in executing tasks at low costs. From environment dynamics point of view, PP methods that can effectively respond to the uncertainties/dynamics emerging from the environment are paramount. For example, in LPP, decision making to avoid obstacles with as least cost as possible can enhance UAV's endurance in the aerial missions. In this regard, LPP methods that can cope up with the

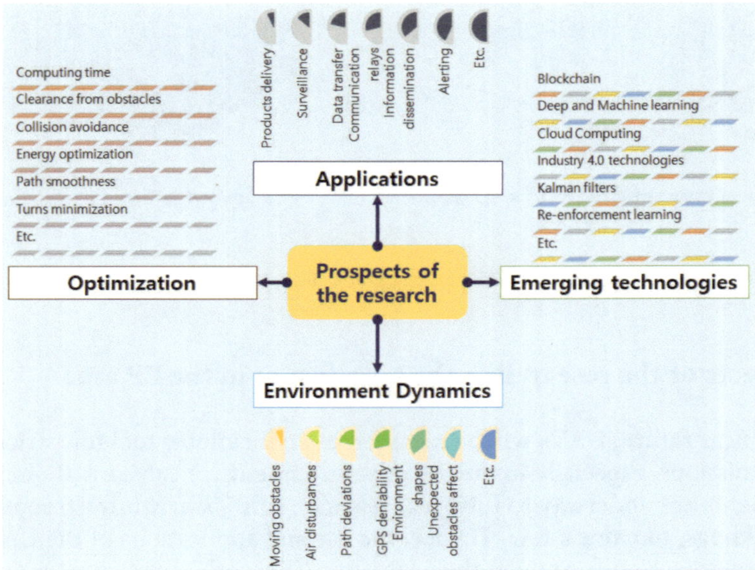

Figure 9.
Categorization of the avenues of future research in the PP/UAVs area.

underlying operating environment variations and can ensure UAV's safety consistently in the practical applications are paramount.

Recently, many emerging technologies have been integrated with the UAV technology. For example, blockchain, transfer learning, computer vision, federated learning, 5G and 6G technologies, and cloud computing etc. have revolutionized the UAVs' applications. In this regard, incorporating more emerging technologies in the UAV domain, and extending the current emerging technologies use to more application areas is an important research direction for the future. Furthermore, improving the hardware capabilities of the UAV by integrating latest technologies are important need from technical perspectives. Despite the technical aspects mentioned above, tailoring computer vision applications in the UAV area is a most promising avenue of the research considering UAV abilities to capture images with good resolution [52]. In addition, identifying niche areas (i.e., water quality analysis, target tracking, covering spatially distributed regions, and detection of wildfire smoke, to name a few) where UAVs can perform well compared to humans, and performing cost–benefit analysis of the UAVs versus human is important research direction in the UAVs' technology. Finally, exploring the possibilities towards joint use of multiple latest technologies in order to serve mankind in an effective way using UAVs is a vibrant area of research. Apart from the PP, devising low-cost CPP methods for UAVs is also an attractive area of research in the near future. Development from hardware perspectives (e.g., battery power, wing-span, payload capabilities, robust decision making abilities, and control aspects) are also a potential avenues for development/research.

6. Conclusions

In this chapter, we have presented concepts, methods, and future research prospects in the area of path planning (PP) for unmanned aerial vehicles (UAVs). Specifically, we have presented the high-level categorization of the PP approaches based on the availability of information regarding UAV operating environment, and UAV strengths. We have discussed three essential components of the PP approaches that are widely adopted by most of the PP approaches. We have discussed substantial number of performance objectives that are improved/optimized by the PP approaches via new concepts/propositions. Furthermore, we have discussed latest approaches that have been proposed to lower the time complexity of pathfinding and their technical challenges. We have described various PP approaches that are used for the PP in unknown environments (aka local PP). We have briefly described the concepts of coverage path planning (CPP) that is subtopic of the PP. The prospects of future research in the UAVs PP area keeping emerging technologies in the loop have also been discussed. With this concise overview, we aim to provide deep understanding about the PP concepts related to the UAVs, and need of the further developments/research in order to enhance UAVs endurance in the airspace specifically in the urban environments. The contents presented in this chapter can help early researchers to quickly grasp the status of existing developments and potential avenues of the research in this area.

Acknowledgements

This work was supported by the National Research Foundation of Korea (NRF) grant funded by the Korea government (MSIT) (2020R1A2B5B01002145).

Conflict of interest

The authors declare no conflict of interest.

Author details

Abdul Majeed*† and Seong Oun Hwang*†
Department of Computer Engineering, Gachon University, Seongnam, South Korea

*Address all correspondence to: ab09@gachon.ac.kr; sohwang@gachon.ac.kr

† These authors contributed equally.

References

[1] Ozkan O, Atli O. Transporting COVID-19 testing specimens by routing unmanned aerial vehicles with range and payload constraints: The case of Istanbul. Transportation Letters. 2021:1-10

[2] Kumar A, Sharma K, Singh H, Naugriya SG, Gill SS, Buyya R. A drone-based networked system and methods for combating coronavirus disease (COVID-19) pandemic. Future Generation Computer Systems. 2021; **115**:1-19

[3] Gao A, Murphy RR, Chen W, Dagnino G, Fischer P, Gutierrez MG, et al. Progress in robotics for combating infectious diseases. Science Robotics. 2021;**52**:6

[4] Vlahogianni EI, Del Ser J, Kepaptsoglou K, Laña I. Model free identification of traffic conditions using unmanned aerial vehicles and deep learning. Journal of Big Data Analytics in Transportation. 2021;**3**(1):1-13

[5] Li T, Liu J, Zhang W, Ni Y, Wang W, Li Z. UAV-human: A large benchmark for human behavior understanding with unmanned aerial vehicles. In: Proceedings of the IEEE/CVF Conference on Computer Vision and Pattern Recognition, Pp. 16266–16275. 2021

[6] Shmelova T, Lazorenko V, Burlaka O. Unmanned aerial vehicles for smart cities: Estimations of urban locality for optimization flights. In: Methods and Applications of Geospatial Technology in Sustainable Urbanism, Pp. 444–477. IGI Global. 2021

[7] Devi M, Maakar SK, Sinwar D, Jangid M, Sangwan P. Applications of flying ad-hoc network during COVID-19 pandemic. In: IOP Conference Series: Materials Science and Engineering, Vol. 1099, no. 1, p. 012005. IOP Publishing. 2021

[8] Algabri M, Mathkour H, Ramdane H, Alsulaiman M. Comparative study of soft computing techniques for mobile robot navigation in an unknown environment. Computers in human behavior. 2015;**50**:42-56

[9] Cui Z, Wang Y. UAV path planning based on multi-layer reinforcement learning technique. IEEE Access. 2021;**9**: 59486-59497

[10] Lu Y, Xue Z, Xia G-S, Zhang L. A survey on vision-based UAV navigation. Geo-spatial information science. 2018; **21**(1):21-32

[11] Couturier A, Akhloufi MA. A review on absolute visual localization for UAV. Robotics and Autonomous Systems. 2021;**135**:103666

[12] Lv T, Zhao C, Bao J. A global path planning algorithm based on bidirectional SVGA. Journal of Robotics. 2017;**2017**

[13] Zhang X, Duan H. An improved constrained differential evolution algorithm for unmanned aerial vehicle global route planning. Applied Soft Computing. 2015;**26**:270-284

[14] Yang X-S. Firefly algorithm, stochastic test functions and design optimisation. International journal of bio-inspired computation. 2010;**2**(2): 78-84

[15] Dorigo M, Maniezzo V, Colorni A. Ant system: Optimization by a colony of cooperating agents. IEEE Transactions on Systems, Man, and Cybernetics, Part B (Cybernetics). 1996;**26**(1):29-41

[16] Roberge V, Tarbouchi M, Labonté G. Comparison of parallel genetic algorithm and particle swarm optimization for real-time UAV path planning. IEEE Transactions on

industrial informatics. 2012;**9**(1): 132-141

[17] Kiran MS, Hakli H, Gunduz M, Uguz H. Artificial bee colony algorithm with variable search strategy for continuous optimization. Information Sciences. 2015;**300**:140-157

[18] Zhang Y, Wang S, Ji G. A comprehensive survey on particle swarm optimization algorithm and its applications. Mathematical Problems in Engineering. 2015;**2015**

[19] Xiang X, Yu C, Lapierre L, Zhang J, Zhang Q. Survey on fuzzy-logic-based guidance and control of marine surface vehicles and underwater vehicles. International Journal of Fuzzy Systems. 2018;**20**(2):572-586

[20] Formato RA. Central force optimization. Prog Electromagn Res. 2007;**77**:425-491

[21] Li P, Duan HB. Path planning of unmanned aerial vehicle based on improved gravitational search algorithm. Science China Technological Sciences. 2012;**55**(10):2712-2719

[22] Meng H, Xin G. UAV route planning based on the genetic simulated annealing algorithm. In: 2010 IEEE International Conference on Mechatronics and Automation, Pp. 788–793. IEEE. 2010

[23] Maini P, Sujit PB. Path planning for a uav with kinematic constraints in the presence of polygonal obstacles. In: 2016 International Conference on Unmanned Aircraft Systems (ICUAS), Pp. 62–67. IEEE. 2016

[24] Frontera G, Martín DJ, Besada JA, Da-Wei G. Approximate 3D Euclidean shortest paths for unmanned aircraft in urban environments. Journal of Intelligent and Robotic Systems. 2017; **85**(2):353-368

[25] Ahmad Z, Ullah F, Tran C, Lee S. Efficient energy flight path planning algorithm using 3-d visibility roadmap for small unmanned aerial vehicle. International Journal of Aerospace Engineering. 2017;**2017**

[26] Majeed A, Lee S. A fast global flight path planning algorithm based on space circumscription and sparse visibility graph for unmanned aerial vehicle. Electronics. 2018;**7**(12):375

[27] Han J. An efficient approach to 3D path planning. Information Sciences. 2019;**478**:318-330

[28] Ghambari S, Lepagnot J, Jourdan L, Idoumghar L. UAV path planning in the presence of static and dynamic obstacles. In: 2020 IEEE Symposium Series on Computational Intelligence (SSCI), Pp. 465–472. IEEE. 2020

[29] Majeed A, Hwang SO. Path planning method for UAVs based on constrained polygonal space and an extremely sparse waypoint graph. Applied Sciences. 2021; **11**(12):5340

[30] Stecz W, Gromada K. UAV mission planning with SAR application. Sensors. 2020;**20**(4):1080

[31] Stecz W, Gromada K. Determining UAV flight trajectory for target recognition using EO/IR and SAR. Sensors. 2020;**20**(19):5712

[32] Siemiatkowska B, Stecz W. A framework for planning and execution of drone swarm missions in a hostile environment. Sensors. 2021;**21**(12):4150

[33] Hong Y, Jung S, Kim S, Cha J. Multi-UAV routing with priority using mixed integer linear programming. In: 2020 20th International Conference on Control, Automation and Systems (ICCAS), Pp. 699–702. IEEE. 2020

[34] Hua X, Wang Z, Yao H, Li B, Shi C, Zuo J. Research on many-to-many target

assignment for unmanned aerial vehicle swarm in three-dimensional scenarios. Computers and Electrical Engineering. 2021;**91**:107067

[35] Cui Z, Wang Y. UAV path planning based on multi-layer reinforcement learning technique. IEEE Access. 2021;**9**: 59486-59497

[36] Maw AA, Tyan M, Nguyen TA, Lee J-W. iADA*-RL: Anytime graph-based path planning with deep reinforcement learning for an autonomous UAV. Applied Sciences. 2021;**11**(9):3948

[37] Wei X, Jianliang X. Distributed path planning of unmanned aerial vehicle communication chain based on dual decomposition. Wireless Communications and Mobile Computing. 2021;**2021**

[38] Zhang, Na, Mingcheng Zhang, and Kin Huat Low. "3D path planning and real-time collision resolution of multirotor drone operations in complex urban low-altitude airspace." Transportation Research Part C: Emerging Technologies 129 (2021): 103123.

[39] Zammit C, Van Kampen E-J. 3D real-time path planning of UAVs in dynamic environments in the presence of uncertainty. In: AIAA Scitech 2021 Forum, p. 1956. 2021

[40] Wu J, Wang H, Zhang M, Yue Y. On obstacle avoidance path planning in unknown 3D environments: A fluid-based framework. ISA transactions. 2021;**111**:249-264

[41] Bayerlein H, Theile M, Caccamo M, Gesbert D. Multi-uav path planning for wireless data harvesting with deep reinforcement learning. IEEE Open Journal of the Communications Society. 2021;**2**:1171-1187

[42] Jamshidi V, Nekoukar V, Refan MH. Real time UAV path planning by parallel grey wolf optimization with align coefficient on CAN bus. Cluster Computing. 2021:1-15

[43] Yan F, Xia E, Li Z, Zhou Z. Sampling-based path planning for high-quality aerial 3D reconstruction of urban scenes. Remote Sensing. 2021; **13**(5):989

[44] Sangeetha V, Krishankumar R, Ravichandran KS, Kar S. Energy-efficient green ant colony optimization for path planning in dynamic 3D environments. Soft Computing. 2021; **25**(6):4749-4769

[45] Sangeetha V, Krishankumar R, Ravichandran KS, Cavallaro F, Kar S, Pamucar D, et al. A fuzzy gain-based dynamic ant Colony optimization for path planning in dynamic environments. Symmetry. 2021;**13**(2): 280

[46] Choi YJ, Rahim T, Nyoman Apraz Ramatryana I, Shin SY. Improved CNN-based path planning for stairs climbing in autonomous UAV with LiDAR sensor. In: 2021 International Conference on Electronics, Information, and Communication (ICEIC), Pp. 1–7. IEEE. 2021

[47] Barnawi A, Chhikara P, Tekchandani R, Kumar N, Alzahrani B. Artificial intelligence-enabled internet of things-based system for COVID-19 screening using aerial thermal imaging. In: Future Generation Computer Systems. 2021

[48] Chen J. Chenglie Du. Ying Zhang: Pengcheng Han, and Wei Wei. "A clustering-based coverage path planning method for autonomous heterogeneous UAVs." IEEE Transactions on Intelligent Transportation Systems; 2021

[49] Majeed A, Lee S. A new coverage flight path planning algorithm based on footprint sweep fitting for unmanned aerial vehicle navigation in urban

environments. Applied Sciences. 2019; **9**(7):1470

[50] Gupta, Rajesh, Aparna Kumari, and Sudeep Tanwar. "Fusion of blockchain and artificial intelligence for secure drone networking underlying 5G communications." Transactions on Emerging Telecommunications Technologies 32, no. 1 (2021): e4176.

[51] Zuhair M, Patel F, Navapara D, Bhattacharya P, Saraswat D. BloCoV6: A blockchain-based 6G-assisted UAV contact tracing scheme for COVID-19 pandemic. In: 2021 2nd International Conference on Intelligent Engineering and Management (ICIEM), Pp. 271–276. IEEE. 2021

[52] Donmez C, Villi O, Berberoglu S, Cilek A. Computer vision-based citrus tree detection in a cultivated environment using UAV imagery. Computers and Electronics in Agriculture. 2021;**187**:106273

Chapter 2

Tracking Control of Unmanned Aerial Vehicle for Power Line Inspection

Kenta Takaya, Hiroshi Ohta, Keishi Shibayama and Valeri Kroumov

Abstract

This work presents some results about power transmission line tracking control and a full autonomous inspection using a quadrotor helicopter. The presented in this paper power line autonomous inspection allows detecting power line defects caused by thunderstorms, corrosion, insulator malfunctions, and same time monitoring of vegetation under the power line corridor. Traditional inspection is performed by helicopters equipped with high-resolution cameras or by direct visual examination carried out by highly skilled staff climbing over de-energized power lines. However, the visual inspection is time-expensive and costly. Moreover, due to regulatory constraints, the helicopters cannot cover narrow mountainous areas. Unmanned aerial vehicles (UAV) are an attractive alternative for power line inspection. In this work, a mathematical model for the quadrotor helicopter used in the autonomous inspection is presented. The model is successfully evaluated through simulations and flight experiments. Next, the construction of a quadrotor helicopter system and its application to power line autonomous inspection is introduced. Simulation and experimental results demonstrate the efficiency and applicability of that system. The results of this research are in the process of implementation for regular inspection of electrical transmission lines.

Keywords: power line tracking, UAV navigation, power line inspection, quadrotor helicopter, field robotics

1. Introduction

Electric power companies worldwide are obliged to guarantee disruptive electrical power supply. The power transmission facilities mainly include power lines, towers, and insulators. These facilities are exposed to thunderstorms, thermal deviations, ice, rain, pollutions like volcanic gases and sour rains: a severe environment that may lead to material fatigue, oxidation, and corrosion. Electric companies are required to inspect and maintain the power transmission equipment periodically. Ground patrols partially inspect these facilities, and, as shown in **Figure 1**, direct visual examination is carried out by skilled personnel climbing over de-energized power lines. As yet, visual inspection is time-expensive and labour-intensive. A common approach nowadays is to use helicopters equipped with high-resolution cameras, but in such inspection, helicopters cannot cover narrow mountainous

Figure 1.
Power line inspection works.

regions due to regulatory constraints. According to the Statistical Report of the Federation of Electric Power Companies in Japan, the total length of high voltage power transmission lines in the country in 2010 is more than 100,000 km, and the number of power transmission towers is much more than 220,000 [1]. Most power transmission facilities in Japan are situated in mountainous regions with no ordinary roads, making the inspection time-consuming and costly.

Unmanned aerial vehicles (UAV) are a promising solution for power line inspection because of time and cost efficiency and because UAVs can approach and inspect energized lines safely. Hence, a significant amount of research is addressing this field. Many researchers have applied computer vision techniques for power transmission towers, and insulators recognition [2–8]. Image processing algorithms are also heavily employed to power lines recognition and tracking [2, 9–17]. However, in most of these studies, distance to the lines is not measured, and their robustness is challenging. In [2] a Region-Based Convolutional Neural Network (R-CNN) is used to localization of transmission towers, and their UAV navigation relies on real-time image processing. The authors of [2] claim that "It is the first time to navigate UAV simultaneously utilizing transmission towers and lines". In [9, 15] after the quadrotor helicopter is navigated to the start point of inspection manually, the vehicle performs autonomous waypoint flight above power lines. Additionally, a Light Detection and Ranging (LiDAR) sensor detect and reconstruct the power line shape [15]. Nevertheless, such navigation may be applied to power distribution lines, but manual flight cannot be performed safely enough when there is a substantial distance to the inspection object. In [11] power transmission lines situated in a highly-populated area are successfully tracked in waypoint flight mode with an aerial speed of 8 m/s. Similar algorithms for power transmission line tracking using position-based visual servo controllers are developed in [10, 12, 13] and are evaluated through several simulations. In [16] an image-based visual servoing combined with a linear quadratic servo control is developed. Deep reinforcement learning is quite successfully applied to autonomous line tracking in [18]. However, network training is performed within a simulation environment, and the robustness of that approach in the real world remains a challenge. The distance to the power line in [18] is measured using a depth camera or stereo camera.

Because of the great significance of automated power lines inspection, power supply companies are rigorously approaching the problem, too [19–23]. However, as far as we know, UAVs are not yet deployed to inspect power line transmission facilities. Hydro-Québec [19] is developing a robot called LineRanger to inspect transmission line conductor bundles. The robot is directly attached to the power line bundle and automatically crosses obstacles like line separators and suspension insulators. LineRanger is equipped with a high-resolution camera and a LiDAR for vegetation monitoring. The system can cover several kilometers a day. They have also developed a similar robotic system called LineScout, capable of inspecting a single transmission line [20]. It can acquire visual information and measure the joints' electrical resistance and monitor the corrosion level of the conductors. DJI M200 Series drones produced by DJI are effectively deployed for power line inspection mainly under manual operation [21]. Although manual vehicle guidance is not time-effective, such inspection is much better than the direct one. The University of KwaZulu-Natal is working with Eskom Holdings SOC Ltd. to develop a power line inspection robot [22], which can climb around jumpers and suspension clamps on a single power transmission line or ground wire. The Electric Power Research Institute (EPRI) is developing a transmission line inspection robot [23], known as "Ti," that can be permanently installed to traverse about 130 km of transmission lines. The robot can transmit in real-time weather data, vegetation imaging, and detect any obstructions on lines. EPRI also worked with manufacturer RADēCO Inc. and Exyn Technologies to develop "an autonomous drone to inspect components in elevated hard-to-access areas, search for temperature anomalies, and collect dose rate surveys in radiological zones" [24]. With UAV's Level 4 technology, the UAV can perform a free-flight exploration of complex spaces while collecting data from the environment [25, 26]. This technology will speed up the implementation of UAVs in the autonomous inspection.

This work presents a novel quadrotor-based system for power line inspection. An outline and some preliminary results of this research are presented in [27], but this paper describes the power line tracking algorithm and its effectiveness and applicability. The main contributions of this study are (1) the development of a quadrotor based system for autonomous inspection of electrical energy transmission and distribution assets. The presented in this paper system has almost Level 4 autonomy in the sense that it can perform flights beyond visual line of sight and without operator based navigation. (2) As far as we are concerned, this is the first time a quadrotor UAV is used in a real application for full autonomous inspection of power lines. (3) The developed system can be applied to almost any industrial multirotor helicopter able to carry the payload of sensors and cameras necessary for inspection.

The rest of the paper is organized as follows. Section 2 presents the quadrotor model used in this development. Section 3 describes the hardware and software configuration of the quadrotor system briefly. The quadrotor model described in Section 2 is evaluated in Section 4 and simulation and experimental results demonstrating the usability of the developed tracking system are depicted there. The last section concludes the paper and gives some plans for further expansion of this work.

2. Quadrotor helicopter dynamics and control

2.1 Quadrotor dynamics

The quadrotor configuration and its reference coordinate systems are shown in **Figure 2**. The vehicle model is acquired under the following assumptions:

1. The quadrotor construction is rigid and symmetrical.

2. The propellers are rigid.

3. The ground effect is neglected, and the centre of gravity of the vehicle lies at the origin of its body coordinate system.

The world reference frame is O_n and the body fixed frame is O_b. The x_b axis of body frame points at normal flight direction, y_b is in starboard side, and z_b axis is in descend direction. The absolute position of the quadrotor body frame $\boldsymbol{p}^n = \left[x_n \ y_n \ z_n\right]^T$ is given by three coordinates of the center of mass in O_n, and its attitude is given by three Euler angles $\boldsymbol{r}^b = [\phi \quad \theta \quad \psi]^T$ standing for roll, pitch and yaw angles, respectively.

The dynamics of the quadrotor helicopter expressed in Newton–Euler notation [28–31] as:

$$\dot{\boldsymbol{p}}^n = \boldsymbol{v}^n \tag{1}$$

$$\dot{\boldsymbol{v}}^n = m^{-1} C_b^n \boldsymbol{F}_b \tag{2}$$

$$\dot{\boldsymbol{r}}^b = \omega^b \tag{3}$$

$$\dot{\boldsymbol{\omega}}^b = \boldsymbol{J}^{-1}\boldsymbol{M}_b + \boldsymbol{J}^{-1}\left(\boldsymbol{\omega}^b \times \boldsymbol{J}\boldsymbol{\omega}^b\right), \tag{4}$$

where m is the mass of the quadrotor frame and C_b^n is the rotation matrix to transform the body frame into world reference frame O_n. The quadrotor translational and rotational motions control is performed by properly changing the thrust F_{T_i}, $i = 1, .., 4$ of rotors. The thrust of the rotors varies by changing their angular speed. The rotor thrust is proportional to the square of the angular rotor speed:

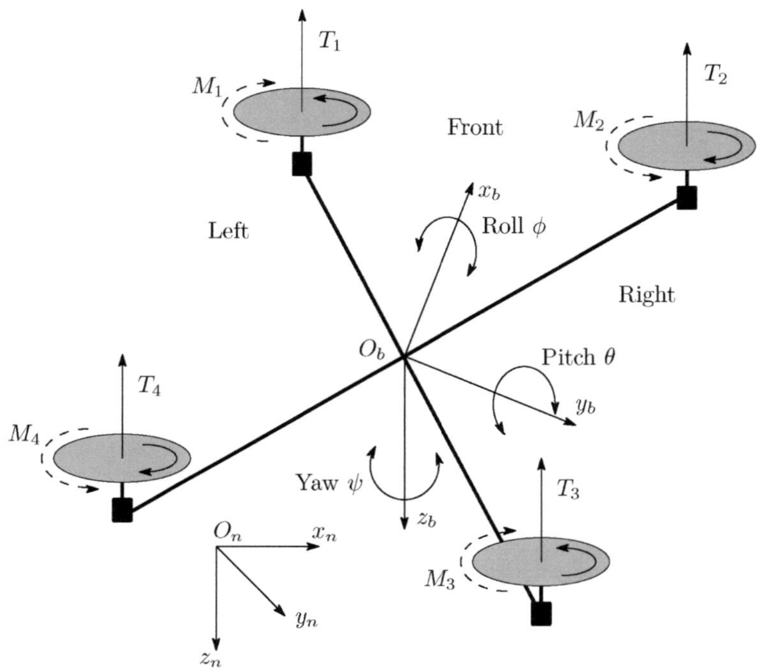

Figure 2.
Quadrotor UAV body and world coordinate systems.

$$F_{T_i} = C_T \omega_{m_i}^2, \tag{5}$$

where C_T is a thrust coefficient, and the angular speed of the rotor is [28]:

$$\dot{\omega}_{m_i} = \frac{1}{J_m}\left(\frac{k_e}{R_m}(U_i - k_e\omega_{m_i}) - (k_t F_{T_i} + D\omega_{m_i})\right), \quad i = 1, .., 4, \tag{6}$$

where U_i is the voltage applied to the motor, J_m is the moment of inertia of the armature and the propeller, R_m is the resistance of the armature coil, k_e and k_t are the electromotive force and torque constants, respectively, and D is the motor viscous friction constant.

The torque generated by the motor i is:

$$M_i = k_t F_{T_i} = k_t C_T \omega_{m_i}^2. \tag{7}$$

The total thrust F_z is:

$$F_z = \sum_{i=1}^{4} F_{T_i}, \tag{8}$$

and control torque vector produced by the four motors is:

$$\boldsymbol{M} = \begin{bmatrix} M_x \\ M_y \\ M_z \end{bmatrix} = \begin{bmatrix} ((F_{T_1} + F_{T_4}) - (F_{T_2} + F_{T_3}))l_m/\sqrt{2} \\ ((F_{T_1} + F_{T_2}) - (F_{T_3} + F_{T_4}))l_m/\sqrt{2} \\ -M_1 + M_2 - M_3 + M_4 \end{bmatrix}. \tag{9}$$

The propeller gyro moments in roll and yaw directions are:

$$M_{jx} = -J_r \omega_r \dot{\theta} \tag{10}$$

$$M_{jy} = J_r \omega_r \dot{\psi} \tag{11}$$

where J_r is the rotor inertia and $\omega_r = \omega_{m_1} - \omega_{m_2} + \omega_{m_3} - \omega_{m_4}$.
The gravitational force in z direction is:

$$F_g = mC_b^n [0 \quad 0 \quad g]^T. \tag{12}$$

The drag reluctant force is a result of the air friction during vehicle movement and rotation:

$$F_d = C_{d,F}|\boldsymbol{v}^b|\boldsymbol{v}^b \tag{13}$$

$$M_d = C_{d,M}|\boldsymbol{\omega}^b|\boldsymbol{\omega}^b \tag{14}$$

where, $C_{d,F}$ and $C_{d,M}$ are aerodynamic coefficient matrices:

$$\begin{aligned} C_{d,F} &= \mathrm{diag}\left[C_{d,F_x} \quad C_{d,F_y} \quad C_{d,F_z}\right] \\ C_{d,M} &= \mathrm{diag}\left[C_{d,M_x} \quad C_{d,M_y} \quad C_{d,M_z}\right]. \end{aligned} \tag{15}$$

The drag generated by the lateral wind is proportional to square of the wind speed:

$$F_w = C_{d,F}|\boldsymbol{v_w}|\boldsymbol{v_w}. \tag{16}$$

The speed dynamics is expressed as:

$$\begin{bmatrix} \dot{v}_x^b \\ \dot{v}_y^b \\ \dot{v}_z^b \end{bmatrix} = \frac{1}{m} \begin{bmatrix} (S\phi S\psi - C\phi S\theta C\psi)F_z - F_{d,x} - F_{w,x} \\ (S\phi C\psi + C\phi S\theta S\psi)F_z - F_{d,y} - F_{w,y} \\ mg - C\phi C\theta F_z - F_{d,z} - F_{w,z} \end{bmatrix}, \tag{17}$$

where C and S stand for sin and cos functions.

Now, for simplicity the gyro angular moment and vehicle angular moment are included in disturbance vector M_d:

$$\begin{bmatrix} \ddot{\phi} \\ \ddot{\theta} \\ \ddot{\psi} \end{bmatrix} = \begin{bmatrix} J_x^{-1}(M_x - M_{d,x}) \\ J_y^{-1}(M_y - M_{d,y}) \\ J_z^{-1}(M_z - M_{d,z}) \end{bmatrix} \tag{18}$$

Finally, the vehicle dynamics, including the motor models, become [28, 30, 32]:

$$\begin{aligned}
\dot{x} &= v_x^b \\
\dot{y} &= v_y^b \\
\dot{z} &= v_z^b \\
\ddot{x} &= m^{-1}((S\phi S\psi - C\phi S\theta C\psi)F_z - F_{d,x} - F_{w,x}) \\
\ddot{y} &= m^{-1}((S\phi S\psi - C\phi S\theta C\psi)F_z - F_{d,y} - F_{w,y}) \\
\ddot{z} &= m^{-1}((mg - C\phi C\theta)F_z - F_{d,z} - F_{w,z}) \\
\dot{\phi} &= \omega_x \\
\dot{\theta} &= \omega_y \\
\dot{\psi} &= \omega_z \\
\ddot{\phi} &= J_x^{-1}\left(((F_{T_1} + F_{T_4}) - (F_{T_2} + F_{T_3}))l_M/\sqrt{2} - M_{d,x}\right) \\
\ddot{\theta} &= J_y^{-1}\left(((F_{T_1} + F_{T_4}) - (F_{T_2} + F_{T_3}))l_M/\sqrt{2} - M_{d,y}\right) \\
\ddot{\psi} &= J_z^{-1}((-M_1 + M_2 - M_3 + M_4) - M_{d,z}) \\
\dot{\omega}_{m_1} &= J_m^{-1}\left(\frac{k_e}{R_m}(U_1 - k_e\omega_{m_1}) - k_t F_{T_1}\right) \\
\dot{\omega}_{m_2} &= J_m^{-1}\left(\frac{k_e}{R_m}(U_2 - k_e\omega_{m_2}) - k_t F_{T_2}\right) \\
\dot{\omega}_{m_3} &= J_m^{-1}\left(\frac{k_e}{R_m}(U_3 - k_e\omega_{m_3}) - k_t F_{T_3}\right) \\
\dot{\omega}_{m_4} &= J_m^{-1}\left(\frac{k_e}{R_m}(U_4 - k_e\omega_{m_4}) - k_t F_{T_4}\right),
\end{aligned} \tag{19}$$

where the viscous friction D in (6) is neglected. The frame control inputs become the motor drive voltages U_i.

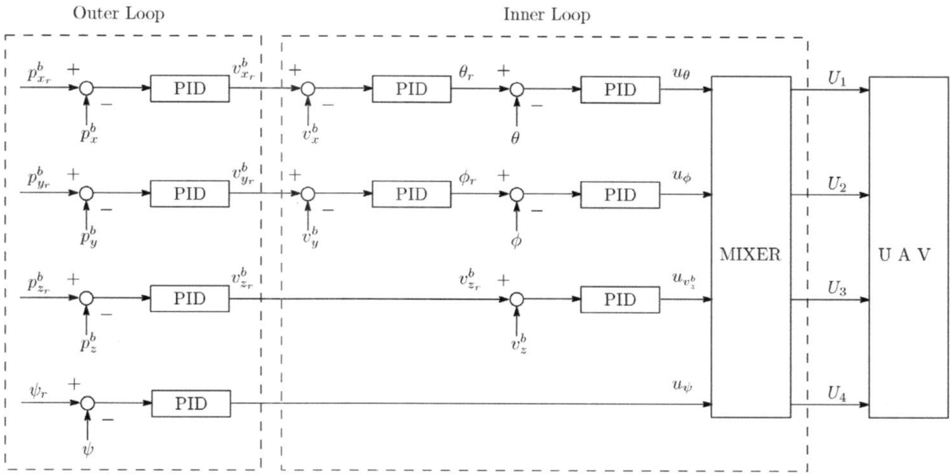

Figure 3.
Quadrotor helicopter cascaded PID control system.

2.2 Control scheme

The overall structure of the closed-loop control of the quadrotor frame is shown in **Figure 3**. The vehicle is an underactuated mechanical system with six degrees of freedom and only four control inputs. The controller consists of a set of cascaded PID controllers arranged in inner and outer loops. The inner loops control the attitude, yaw rate, and vertical velocity. The outer loop controls the altitude and heading. Both loops are designed similarly to those implemented in the actual quadrotor flight controller [33]. This control technique is called a time-scale separation [34] and works well when the inner loops are significantly faster than the outer ones. The MIXER block in **Figure 3** forms the voltages applied to the motors:

$$\begin{aligned} U_1 &= -u_\phi + u_\theta - u_\psi + u_{v_z^b} \\ U_2 &= u_\phi + u_\theta + u_\psi + u_{v_z^b} \\ U_3 &= u_\phi - u_\theta - u_\psi + u_{v_z^b} \\ U_4 &= -u_\phi - u_\theta + u_\psi + u_{v_z^b}. \end{aligned} \tag{20}$$

3. Power line autonomous inspection system architecture

3.1 Hardware design

The hardware architecture of the quadrotor system is shown in **Figure 4a**. The power transmission lines are detected using a LiDAR pointing in a vertical direction. Measured distance to the power line by the same LiDAR is used to adjust the zoom and focus of the image acquisition camera. The obstacle avoidance algorithm uses an infrared Time-of-Flight (ToF) sensor having 360° range and pointing horizontally. Because the magnetic field generated by the power lines interferes with the internal compass of the flight controller, the heading of the vehicle is estimated through a moving baseline real-time kinematics (RTK) [35] technique. For safety reasons, a second GNSS module is added to the frame. A second camera pointing in the down direction performs the vegetation monitoring under the power line corridor. All sensors are connected to NVIDIA Jetson TX2 [36] companion computer, which controls the position of the vehicle via MAVLink protocol [37]. The real-time

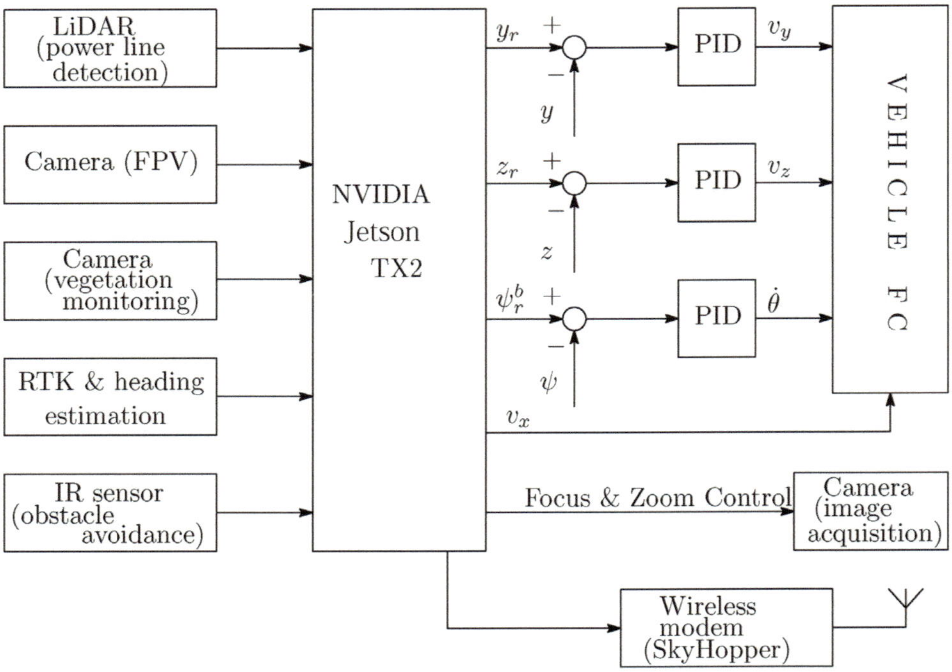

(a)

(b)

Figure 4.
Hardware architecture and vehicle appearance. (a) Hardware architecture, (b) appearance of the quadrotor system.

images from a first-person view (FPV) camera and the complete status of the quadrotor (position in the world and the local coordinate systems, battery level, GNSS status, magnetic field strength) are transmitted to the ground station computer in 1 s intervals for system monitoring. The wireless modem (SkyHopper PRO V [38]) performs all necessary communications between the quadrotor frame and the ground station. The aircraft is stabilized by a Pixhawk flight controller (FC) [39]. However, the altitude z^b and lateral control x^b during line tracking is

Parameter	Value
Propeller diameter [in]	19
Maximum payload [kg]	2.0
Weight [kg]	6.5
Flight time [min]	20
LiDAR resolution [deg]	0.125
RTK module	Yes
Companion computer	Jetson TX2 [36]
Wireless modem	SkyHopper PRO V [38]
Flight controller	Pixhawk [39]

Table 1.
Quadrotor vehicle specifications.

performed using three outer loops realized on the companion computer as shown in **Figure 4a**. FC controls the forward speed with a reference command from the companion computer. The vision system estimates the yaw angle ψ. **Figure 4b** depicts the UAV appearance. The specifications of the quadrotor helicopter are given in **Table 1**.

3.2 Software

The control software is realized in Python language, and most of it runs under Robot Operating System (ROS) [40]. The software consists of modules for sensor data acquisition and processing, camera control node, and node for sensor data transmission to the control process and the same time to the base station computer. The control process receives power line position measured by the LiDAR pointing in the down position in the body coordinate system, GNSS coordinates of the tower to be inspected, GNSS receiver status, battery status, and FC status. The control process uses three PID controllers for altitude and yaw control, as shown in **Figure 4a**. It performs path generation as well as almost all necessary processing for realizing a safe flight.

Even though the vehicle control software communicates with the sensor relating nodes, it runs thoroughly independent from the ROS processes for safety reasons. Because of this autonomy, the control process can bring the vehicle safely back to the home position even if the whole ROS crashes or all sensors except one of the GPS modules get out of order.

3.3 Power line tracking

The inspection process includes the following actions. As shown in **Figure 5**, after taking off at safe altitude, the vehicle approaches the start tower of the inspection path using *a priori* information about the tower world coordinates and its height. Then the electrical line subject to inspection is detected by the LiDAR pointing in a downward direction. Next, the vehicle is positioned above the wire at a distance allowing it to acquire good quality images. While keeping a constant distance to the power line, the quadrotor moves along it at a constant speed and acquires high-resolution images of the wire and the vegetation below. After reaching the last tower subject of inspection, the quadrotor returns safely to the home position.

The autonomous inspection consists of the following steps:

1. Selection of the inspection target using a graphical user interface (GUI) running on the base station computer (see **Figure 6**).

2. Using a priori data about the tower position in the world coordinate system and its mechanical construction, the vehicle takes off to a safe altitude. It approaches the top of the tower using GNSS measurements.

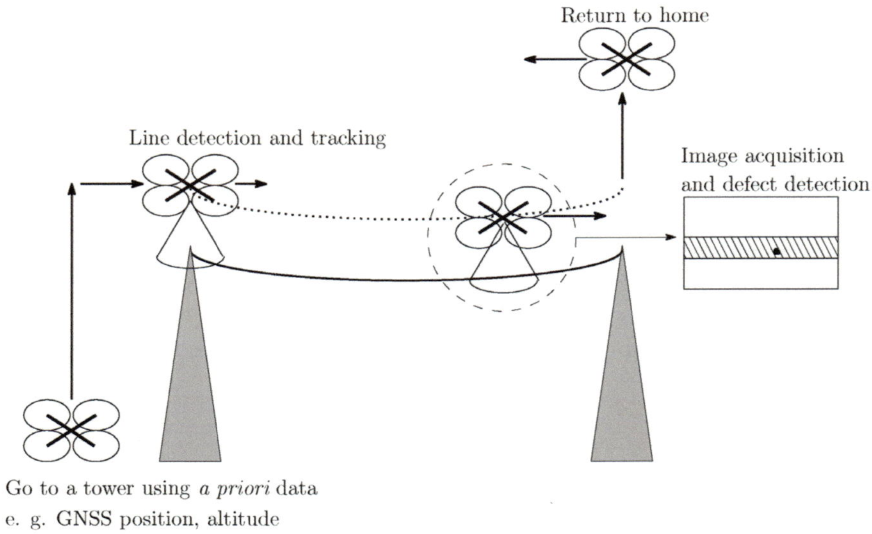

Figure 5.
Power line tracking concept.

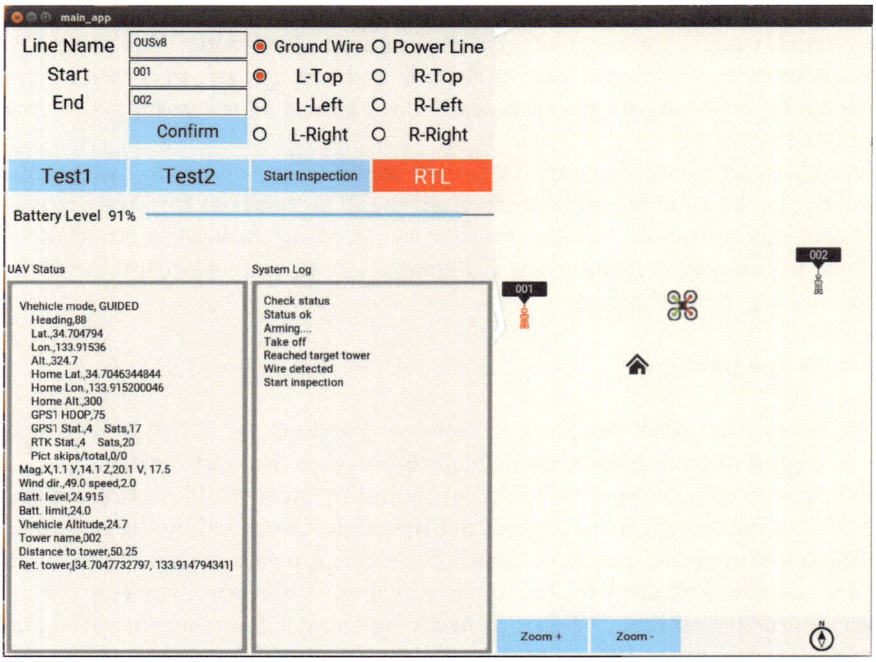

Figure 6.
GUI for power line inspection. GW: ground wire, L: left, R: right, RTL: return to launch.

Figure 7.
Quadrotor system during power line autonomous inspection.

3. Search of the power line subject of the inspection. The search is performed using LiDAR readings of the power line position.

4. Power line tracking control at a certain velocity and constant distance, based on the LiDAR measurements and same time performing continuous image acquisition and image processing for detection of rust and defects on the power line. An additional camera monitors the vegetation under the power line corridor.

5. After finishing the inspection, the UAV returns safely back to the home position.

The GUI is used to send the inspection subject to the quadrotor companion computer before the beginning of the inspection. It also displays the UAV status during the flight and its position on the map. The power transmission towers are displayed on the same map, too. Preflight safety tests of the vehicle and the control software can also be invoked from the same GUI. After the beginning of the inspection, communication from the aircraft only is performed. Only the Return to Launch (RTL) command can be transmitted from the GUI.

The GUI in **Figure 6** shows a setting for a ground wire (GW) inspection. Images of the ground wire can be acquired from above and from a slant direction. The subject shown in the figure is a transmission tower with two ground wires. The left-side wire is selected, and the images from the top above are to be acquired. Left or right side GWs can also be inspected from slant direction: "R-Left" means right-side wire to be scanned from the left-side, "R-Right" means right-side wire to be scanned from the right-side and so on.

Figure 7 shows the UAV during autonomous ground wire inspection. The image in **Figure 7** is acquired by a camera mounted on a second aircraft.

4. Simulation and experimental results

4.1 Step response simulation and experiments

In this section, the quadrotor model (Eq. (19)) is validated by simulation and experiments. The UAV specification is shown in **Table 1**. The parameters of the quadrotor are shown in **Table 2** and PID controller parameters are depicted in **Table 3**. In these experiments, the model and actual vehicle are moved along body coordinates system axes at a constant speed. The simulation and experimental responses are compared in **Figure 8**. **Figure 8a**, **c**, and **e** show the responses during up-down, forward-backward, and left–right flights. The simulation and experimental results are very similar, showing that the model presents the actual vehicle's behavior well. It can be seen that the actual vehicle in up-down flights tends to increase the altitude slightly, and we suppose that might exist some problem in tuning the flight controller filter parameters, causing error in the estimation of the

Parameter	Value	Description
U[V]	25.0	Maximum input voltage
R_m [Ω]	0.013	Armature resistance
k_e [s/rad]	0.0306	Back EMF coefficient
k_t	1.0×10^{-4}	Torque coefficient
J_m [kg·m^2]	6.68×10^{-4}	Rotor moment of inertia
C_{T0} [N/rad^2]	3.37×10^{-5}	Thrust coefficient
m [kg]	7.0	Total weight
l_m [m]	0.275	Motor distance to the center of UAV
J_x [kg· m^2]	0.637	Moment of inertia (x-axis)
J_y [kg· m^2]	0.637	Moment of inertia (y-axis)
J_z [kg· m^2]	0.52	Moment of inertia (z-axis)
C_{d,F_x}	0.167	Drag coefficient (x-axis)
C_{d,F_y}	0.167	Drag coefficient (y-axis)
C_{d,F_z}	0.059	Drag coefficient (z-axis)
C_{d,M_x}	0.059	Rotational drag coefficient (x-axis)
C_{d,M_y}	0.059	Rotational drag coefficient (y-axis)
C_{d,M_z}	0.167	Rotational drag coefficient (z-axis)

Table 2.
UAV parameters.

Controller	P	I	D
Speed v_x^b	0.225	0.225	0.02
Speed v_y^b	0.225	0.225	0.02
Speed v_z^b	25.0	5.0	0.0
Roll ϕ	70.0	25.0	10.0
Pitch θ	70.0	25.0	10.0
Yaw ψ	100.0	3.5	150
Position y^b	0.15	0.0	0.05
Position z^b	0.5	0.03	0.05

Table 3.
PID controllers parameters.

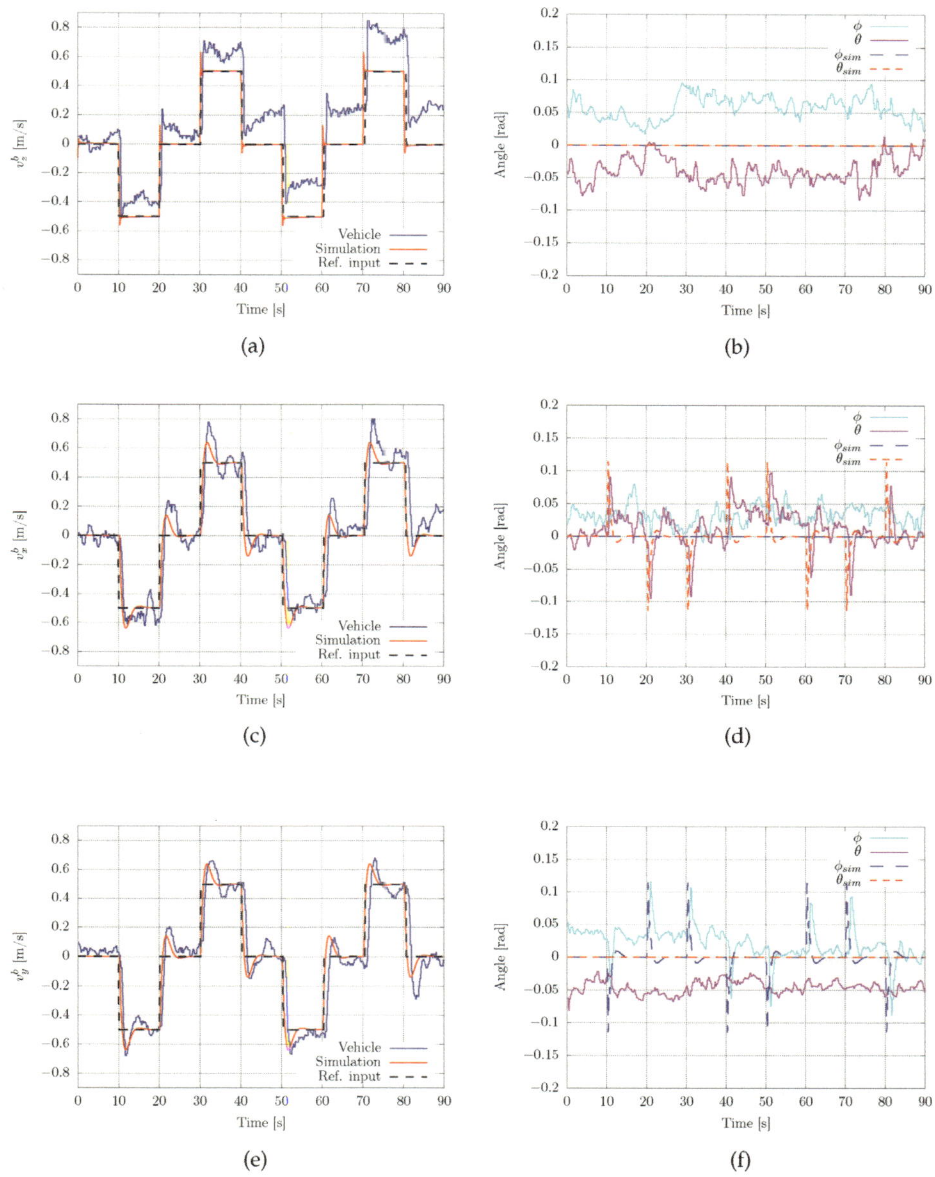

Figure 8.
Step response simulations and experiment results. (a) Up-down movement, (b) roll and pitch angles during up-down flight, (c) flight in forward and backward directions, (d) roll and pitch during forward and backward flight, (e) left and right movement, (f) roll and pitch during left and right flight.

altitude. In **Figure 8b**, **d**, and **f** the body roll and pitch angles are drawn. It can be observed that angles ϕ and θ during simulation and experiments slightly differ. Therefore, we can conclude that the experimental results demonstrate the theoretical expectations.

4.2 Power line tracking simulation

In this section, simulation and experimental results during power line tracking are presented. The simulation setting is given in **Figure 9**. The simulation conditions are as follows:

- At the beginning of the simulation, the initial position of the UAV is 6 m above and 1 m on the port side of the power line.
- After position adjustment above the line for 10 s is done, the vehicle performs a wire tracking flight.
- The position of the power line in the vehicle coordinate system during tracking is set to $z = 3$ m and $y = 0$ m.

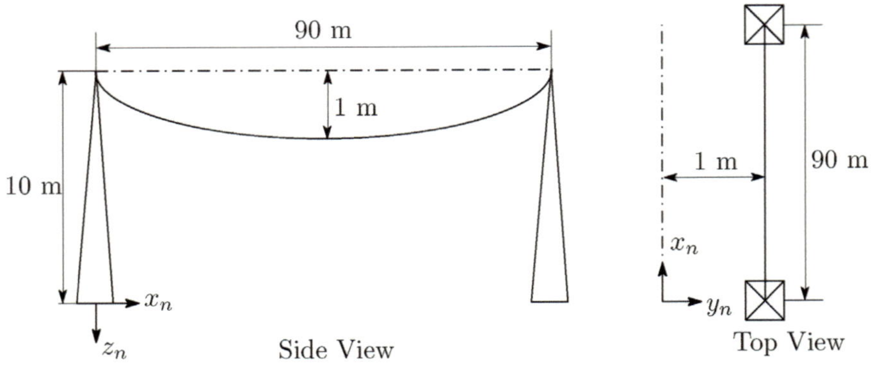

Figure 9.
Setup for power line tracking simulations and experiments.

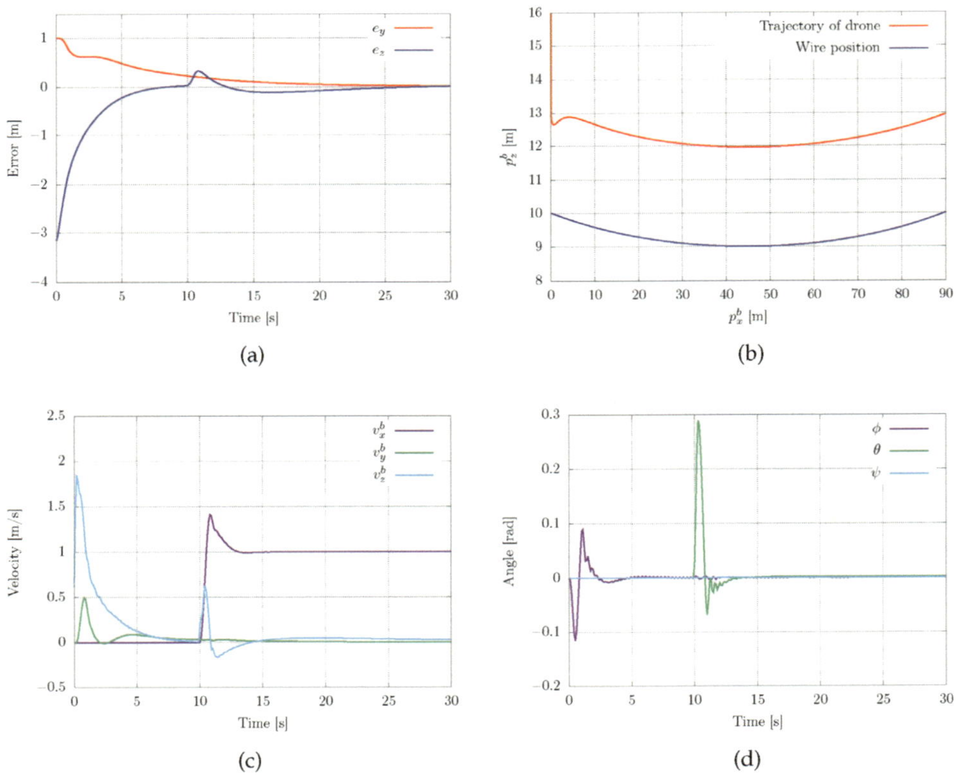

Figure 10.
Power line tracking simulation result. (a) Position error, (b) vehicle position vs power line, (c) vehicle speed, (d) roll and pitch body angles.

To achieve good performance during the simulation, the control frequency of inner loops is 400 Hz while the outer loops are at 10 Hz (see Section 2.2).

In the simulations, the catenary shape of the power line is expressed as:

$$w_z = \cosh\left(a_w(x_n - w_l/2)\right) + w_h, \tag{21}$$

where w_z is the vertical position of the power line in local coordinates, $w_l = 90$ is the distance between two towers the wire is attached to, and taking $a_w = 0.03$, $w_h = 8$ give a sag of 1 m with towers height of 10 m.

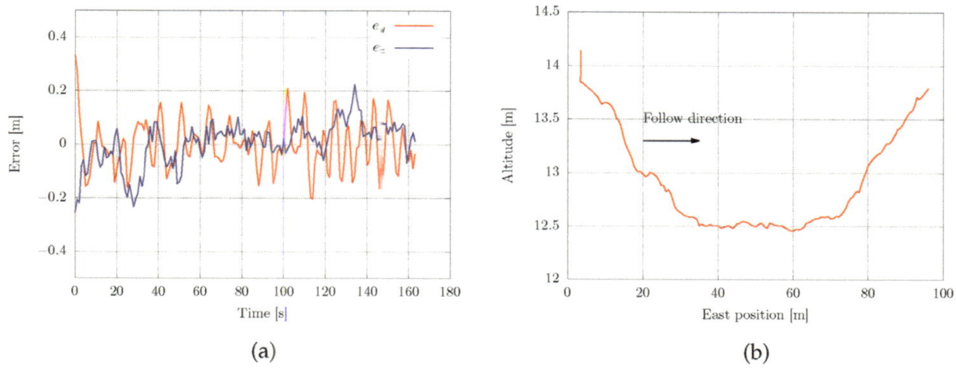

Figure 11.
Ground wire tracking experiment. (a) Position error, (b) UAV path.

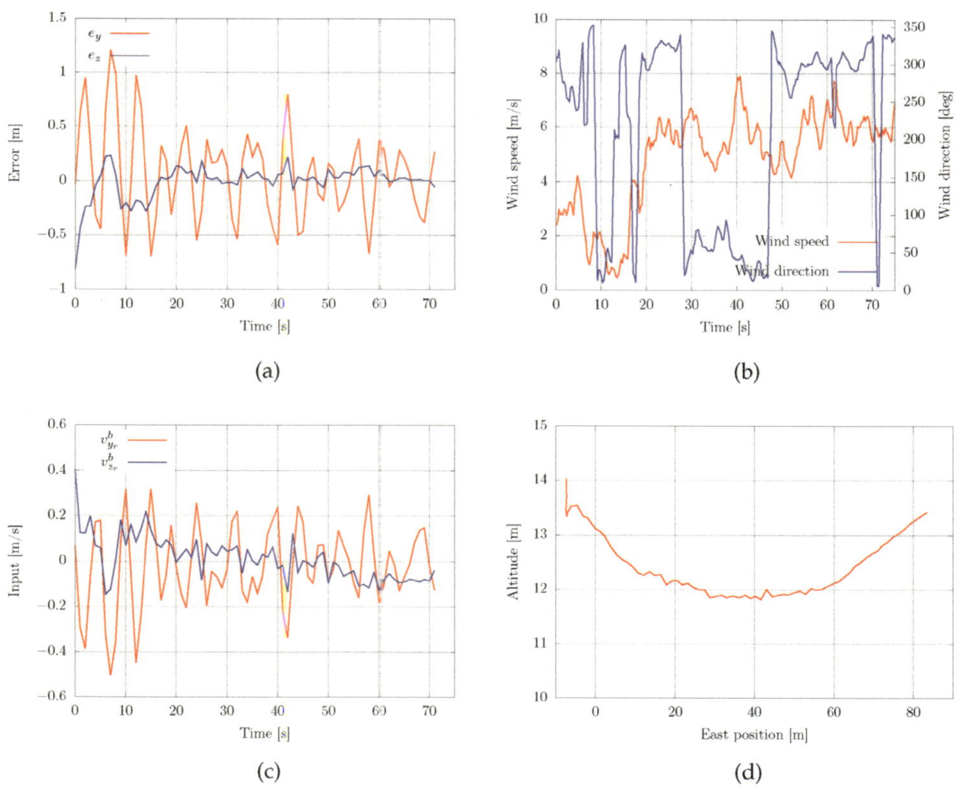

Figure 12.
GW tracking experiment in presence of strong wind. (a) Position error, (b) wind speed and direction, (c) control input, (d) UAV path.

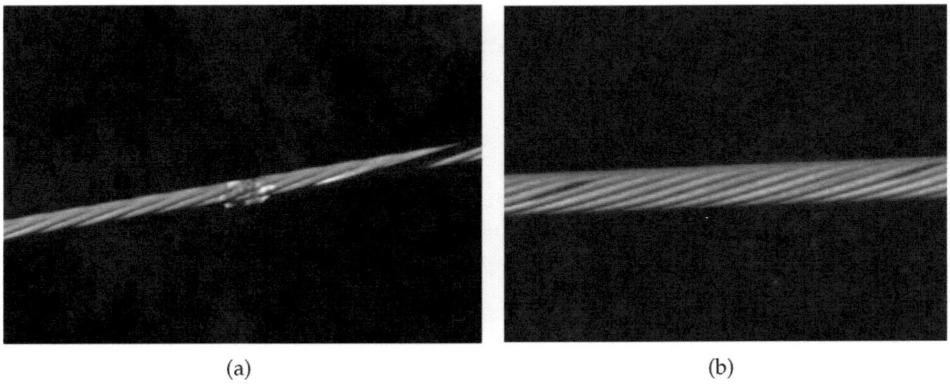

Figure 13.
Images of ground wire. (a) Defect caused by a lightning strike. (b) Two minor defects.

Simulation results are shown in **Figure 10**. It can be seen from the same figure that the position error is negligible and converges quite fast.

4.3 Ground wire tracking experiment

In this section, the experimental results of actual ground line tracking flight are shown. The experimental setting is the same as the simulation one (see **Figure 9**). The diameter of the wire is 8 mm. The flights were performed autonomously, as explained in Section 3.3. The experimental results are shown in **Figures 11** and **12**.

The graphs in **Figure 11a** depict the position errors. The error in the vertical direction is less than ± 0.2 m and but the control quality is good. **Figure 11b** shows the vehicle trajectory—the catenary shape of the wire is well tracked. The vehicle speed was 0.7 m/s.

The graphs in **Figure 12** are taken in the presence of quite strong wind. It can be seen from the graph shown in **Figure 12a** that the position error in a vertical direction is small, but the error in the horizontal direction is about ± 0.5 m. **Figure 12b** shows the wind speed and direction measured using an anemometer mounted on the vehicle body. The wind speed and direction are measured in 0.33 s interval, and graphs present 1 s average values. However, there were sudden changes in the wind speed up to 9 m/s. **Figure 12c** and **d** depict the control input and the catenary shape of vehicle trajectory. The vehicle speed in this experiment was 1.4 m/s.

Two images acquired during vehicle flight are depicted in **Figure 13**. Some minor defects can be observed there.

5. Conclusions

This work aims to develop a reliable autonomous power line tracking and inspection system based on a quadrotor helicopter. The model of the UAV was presented and evaluated in simulation and experiments performed in the real environment. Classical PID controllers were deployed, and their performances were demonstrated during ground wire line tracking. It can be concluded that the PID controller had a good performance, but in windy weather conditions, the position error increases to some extend. The presented system has almost Level 4 autonomy in the sense that it can perform flights beyond visual line of sight and without operator based navigation.

As a part of future work, we will implement a fuzzy PID controller to explore if further performance improvement can be achieved. This system will be implemented on regular inspections and maintenance of power facilities in an electric power company in Japan.

Acknowledgements

This research is performed in cooperation with Electric Power Development Co., LTD. from Tokyo, Japan. The authors would like to thank for their financial support of this project. We want to thank Mr. K. Tanaka, Mr. T. Sugiyama and Mr. Y. Oota from the same company for their help and support in performing most of the experiments.

Author details

Kenta Takaya[1†], Hiroshi Ohta[2†], Keishi Shibayama[3†] and Valeri Kroumov[2*†]

1 Graduate School, Okayama University of Science, Okayama, Japan

2 Okayama University of Science, Okayama, Japan

3 Scale Corporation, Asago City, Hyogo, Japan

*Address all correspondence to: val@ee.ous.ac.jp

† These authors contributed equally.

References

[1] 60 Years Power Industry in Japan: Statistics. https://www.fepc.or.jp/library/data/60tokei/index.html, [Accessed: 02 Jul 2021], In Japanese.

[2] Bian J., Hui X., Zhao X., and Tan M. A novel monocular-based navigation approach for uav autonomous transmission-line inspection. In *2018 IEEE/RSJ International Conference on Intelligent Robots and Systems (IROS)*, pages 1–7, 2018. doi: 10.1109/IROS.2018.8593926.

[3] Yue L., Jun Y., Liang L., Jinlong Z., and Zongyu L. The method of insulator recognition based on deep learning. In *2016 4th International Conference on Applied Robotics for the Power Industry (CARPI)*, pages 1–5, 2016. doi: 10.1109/CARPI.2016.7745630.

[4] Jabid T. and Uddin Md. Z. Rotation invariant power line insulator detection using local directional pattern and support vector machine. In *2016 International Conference on Innovations in Science, Engineering and Technology (ICISET)*, pages 1–4, 2016. doi: 10.1109/ICISET.2016.7856522.

[5] Ohta H., Sato Y., Mori T., Takaya K., and Kroumov V. Image acquisition of power line transmission towers using uav and deep learning technique for insulators localization and recognition. In *2019 23rd International Conference on System Theory, Control and Computing (ICSTCC)*, pages 125–130, 2019. doi: 10.1109/ICSTCC.2019.8885695.

[6] Zhai Y., Chen R., Yang Q., Li X., and Zhao Z. Insulator fault detection based on spatial morphological features of aerial images. IEEE Access, 6: 35316–35326, 2018. doi: 10.1109/ACCESS.2018.2846293.

[7] Tao X., Zhang D., Wang Z., Liu X., Zhang H., and Xu D. Detection of power line insulator defects using aerial images analyzed with convolutional neural networks. IEEE Transactions on Systems, Man, and Cybernetics: Systems, 50(4):1486–1498, 2020. doi: 10.1109/TSMC.2018.2871750.

[8] Jiang H., Qiu X., Chen J., Liu X., Miao X., and S. Zhuang. Insulator fault detection in aerial images based on ensemble learning with multi-level perception. IEEE Access, 7:61797–61810, 2019. doi: 10.1109/ACCESS.2019.2915985.

[9] Luque-Vega L. F., Castillo-Toledo B., Loukianov A., and Gonzalez-Jimenez L. E. Power line inspection via an unmanned aerial system based on the quadrotor helicopter. In *MELECON 2014 – 2014 17th IEEE Mediterranean Electrotechnical Conference*, pages 393–397, 2014. doi: 10.1109/MELCON.2014.6820566.

[10] Li Z., Liu Y., Hayward R., Zhang J., and Cai J. Knowledge-based power line detection for uav surveillance and inspection systems. In *2008 23rd International Conference Image and Vision Computing New Zealand*, pages 1–6, 2008. doi: 10.1109/IVCNZ.2008.4762118.

[11] Zhou G., Yuan J., Yen I-L., and Bastani F. Robust real-time uav based power line detection and tracking. In *2016 IEEE International Conference on Image Processing (ICIP)*, pages 744–748, 2016. doi: 10.1109/ICIP.2016.7532456.

[12] Golightly I. and Jones D. Visual control of an unmanned aerial vehicle for power line inspection. In ICAR '05. *Proceedings., 12th International Conference on Advanced Robotics, 2005.*, pages 288–295, 2005. doi: 10.1109/ICAR.2005.1507426.

[13] Jones D., Golightly I., Roberts J., and Usher K. Modeling and control of a robotic power line inspection vehicle.

In *2006 IEEE Conference on Computer Aided Control System Design, 2006 IEEE International Conference on Control Applications, 2006 IEEE International Symposium on Intelligent Control*, pages 632–637, 2006. doi: 10.1109/CACSD-CCA-ISIC.2006.4776719.

[14] Feng T., Wang Y., and Zhu L. Power line recognition and tracking method for uavs inspection. In *2015 IEEE International Conference on Information and Automation*, pages 2136–2141, 2015. doi: 10.1109/ICInfA.2015.7279641.

[15] Chuang Deng, Jun-yong Liu, You-bo Liu, and Yang-yang Tan. Real time autonomous transmission line following system for quadrotor helicopters. In *2016 International Conference on Smart Grid and Clean Energy Technologies (ICSGCE)*, pages 61–64, 2016. doi: 10.1109/ICSGCE.2016.7876026.

[16] Araar O. and Aouf N. Visual servoing of a quadrotor uav for autonomous power lines inspection. In *22nd Mediterranean Conference on Control and Automation*, pages 1418–1424, 2014. doi: 10.1109/MED.2014.6961575.

[17] Valipour S., Khandani H., and Moradi H. The design and implementation of a hotline tracking uav. In *2015 3rd RSI International Conference on Robotics and Mechatronics (ICROM)*, pages 252–258, 2015. doi: 10.1109/ICRoM.2015.7367793.

[18] Pienroj P., Schönborn S., and Birke R. Exploring deep reinforcement learning for autonomous powerline tracking. In *IEEE INFOCOM 2019 - IEEE Conference on Computer Communications Workshops (INFOCOM WKSHPS)*, pages 496–501, 2019. doi: 10.1109/INFCOMW.2019.8845212.

[19] LineRanger : A Revolution in Transmission Line Robotics. http://www.hydroquebec.com/robots/en/, [Accessed: 08 Aug 2021].

[20] NYPA TESTS ROBOTIC TRANSMISSION INSPECTION TECHNOLOGY. https://www.youtube.com/watch?v=DuKjnaiBP4c, [Accessed: 08 Aug 2021].

[21] DJI – M200 Series– Powerline Inspection Tool. https://www.youtube.com/watch?v=Bj5ByZKVNao, [Accessed: 08 Aug 2021].

[22] Power Line Inspection Robot: Inspection Sample. https://www.youtube.com/watch?v=cTPooHx6rr8, [Accessed: 08 Aug 2021].

[23] Barker B. Robots take on tough tasks in transmission and distribution. *EPRI Journal*, pages 1–6, 02 2019.

[24] EPRI demonstrates autonomous drone. https://www.neimagazine.com/features/featureepri-demonstrates-autonomous-drone-7967221/, [Accessed: 08 Aug 2021].

[25] Explore the most dangerous places with autonomous drones. https://www.exyn.com/, [Accessed: 08 Aug 2021].

[26] Japan's first on-demand drone delivery service for hot lunches to healthcare professionals. https://aeronext.com/news/yokosuka beefbowldelivery/, [Accessed: 08 Aug 2021].

[27] Takaya K., Ohta H., Kroumov V., Shibayama K., and Nakamura M. Development of uav system for autonomous power line inspection. In *2019 23rd International Conference on System Theory, Control and Computing (ICSTCC)*, pages 762–767, 2019. doi: 10.1109/ICSTCC.2019.8885596.

[28] Meyer J., Sendobry A., Kohlbrecher S., and Klingauf U. von Stryk O. Comprehensive simulation of quadrotor uavs using ros and gazebo. In *3rd Int. Conf. on Simulation, Modeling and Programming for Autonomous Robots (SIMPAR)*, pages 400–411, 2012. doi: 10.1007/978-3-642-34327-8 36.

[29] Bouabdallah S. *Design and Control of quadrotors with application to autonomous flying*. PhD thesis, EPFL, Lausanne, 01 2007.

[30] Hoffmann G., Waslander S., and Tomlin C. Quadrotor helicopter trajectory tracking control. In *AIAA Guidance, Navigation and Control Conference and Exhibit*, 2018. doi: 10.2514/6.2008-7410.

[31] Dydek Z. T., Annaswamy A. M., and Lavretsky E. Adaptive control of quadrotor uavs: A design trade study with flight evaluations. IEEE Transactions on Control Systems Technology, 21(4):1400–1406, 2013. doi: 10.1109/TCST.2012.2200104.

[32] Bouabdallah S., Noth A., and Siegwart R. Pid vs lq control techniques applied to an indoor micro quadrotor. In *Proceedings of 2004 IEEE/RSJ International Conference on Intelligent Robots and Systems*, volume 3, pages 2451 – 2456, 01 2004. ISBN 0-7803-8463-6. doi: 10.1109/IROS.2004.1389776.

[33] PX4 Development Guide: PX4 Architectural Overview. https://dev.px4.io/en/concept/architecture.html, [Accessed: 04 May 2021].

[34] Brescianini D., Hehn M., and D'Andrea R. Nonlinear quadrocopter attitude control. technical report. Technical report, EPFL, Zürich, 2013.

[35] GPS for Yaw (aka Moving Baseline). https://ardupilot.org/copter/docs/common-gps-for-yaw.html, [Accessed: 07/06/2021].

[36] Jetson-TX2 (Embedded computer). https://developer.nvidia.com/embedded/jetson-tx2, [Accessed: 04 May 2021].

[37] MAVLink (Micro Air Vehicle Communication Protocol). https://mavlink.io/en/, [Accessed: 04 May 2021].

[38] SkyHopper PRO. https://www.skyhopper.biz/products/communication-data-links/, [Accessed: 04 Jul 2021].

[39] PixHawk (Flight Controller). https://ardupilot.org/copter/docs/common-pixhawk-overview.html, [Accessed: 04 May 2021].

[40] ROS Kinetic Kame. http://wiki.ros.org/kinetic, [Accessed: 04 Jun 2021].

Section 2

AI-Based Optimization for Planning

Chapter 3

The Relationship between "C-Space", "Heuristic Methods", and "Sampling Based Planner"

Emanuele Sansebastiano and Angel P. del Pobil

Abstract

Defining the collision-free C-space is crucial in robotics to find whether a robot can successfully perform a motion. However, the complexity of defining this space increases according to the robot's degree of freedom and the number of obstacles. Heuristics techniques, such as Monte Carlo's simulation, help developers address this problem and speed up the whole process. Many well-known motion planning algorithms, such as RRT, base their popularity on their ability to find sufficiently good representations of the collision-free C-space very quickly by exploiting heuristics methods, but this mathematical relationship is not highlighted in most textbooks and publications. Each book focuses the attention of the reader on C-space at the beginning, but this concept is left behind page after page. Moreover, even though heuristics methods are widely used to boost algorithms, they are never formalized as part of the Optimization techniques subject. The major goal of this chapter is to highlight the mathematical and intuitive relationship between C-space, heuristic methods, and sampling based planner.

Keywords: motion planning, C-space, optimization techniques, heuristic methods, sampling based planner, educational dissemination

1. Introduction

Starting from the concept of configuration space (known as c-map as well), this chapter highlights the necessity of finding an alternative way to perform path search than computing the whole configuration space deterministically. Even simple 2D scenarios appears to be quite difficult to be handled by calculating the whole configuration space.

This chapter comes from the necessity of recalling the fact that such powerful and well-known algorithms like sampling based algorithms are strictly connected to the configuration space. Configuration space is the most ancient concept root of motion planning. Moreover, motion planning algorithms are generally imagined by the people as deterministic analysis of the environment because humans expect robots to be foolproof. Sampling based planners, instead, are coming from heuristic approaches which are the exact opposite of deterministic analysis. Most of the disseminating books and papers [1–6] are focused on explaining the algorithms rather then report how and why those algorithms were born. The main contribution of this chapter is reporting how and why sampling based planners algorithms were born and which challenges older planning algorithms could not solve. Knowing the history and the ideas behind each planning algorithm allows scientists to fully understand its capabilities. Moreover, it allows scientists to know which algorithm is more eligible for a specific problem a priori without testing all of them.

Section 2 explains how to compute a full configuration space and reports several examples in order to fully address the practical meaning of c-map. Moreover, this section describes accurately which are the major drawbacks of computing the whole c-map. In many cases, computing the c-map is practically impossible.

Section 3 describes heuristic methods and their ability to tackle very convolute or large problems quickly without losing much information. Those methods do not ensure any result because they are not deterministic, but they do find a very good solution if they are well designed. This chapter is not reporting any specific heuristic method because they must be designed according to the problem.

Section 4 gives a quick overview of sampling based planners. Sampling based planes comes from the necessity of exploring configuration spaces by using heuristic methods. In particular, this section reports two of the most famous sampling based planes: Probabilistic road map (PRM) and Rapidly-exploring random tree (RRT). Those planners or a variation of them are probably the most use all around the world.

2. Configuration space (C-space)

The parameters which define the configuration of a system are called generalized coordinates, and the vector space defined by these coordinates is called configuration space of the system. The configuration space represents all possible states that the system might have. According to the number of degrees of freedom of a system, its configuration space has to be defined by a vector having the same length. Due to this fact, the representation of configuration spaces on paper is limited to systems having 2 degrees of freedom.

Configuration spaces are often used in robotics to represents all possible mechanical configurations which a robot can have. Moreover, configuration spaces are used to motion planning by stacking several configurations into one motion. Given a starting configuration and a goal configuration there are infinite possible motions which lead the robot from the stating point to the goal one. Assuming that the path search is restricted to geometrical constraints (no kinematics and no dynamics is involved), the cleverest path would be defined by a straight line connecting the starting point and the goal point on the configuration space (**Figure 1**). However, path planning algorithms are mostly used to carry robot through obstacles. Assuming that obstacles are not moving or that they are moving slowly enough to be considered static comparted to the robot motion speed. Configuration space can be computed to describe all possible robot configurations not colliding with any obstacle. These configuration spaces are called "*collision free c-space*".

Creating a configuration space is simple if there are no obstacles in the scene. Knowing the range of values that each degree of freedom can have is enough to build a fully function configuration space map. Including obstacles into the scene increases significantly the complexity of building configuration space map.

Before continuing this chapter, it is better to highlight that in order to simplify the handling of the topic and help the reader to visualize all examples, all presented robots and obstacles lay into a 2D plane. Moreover, all robots are going to have just 2 degrees of freedom or less to allow 2D representation of configuration spaces.

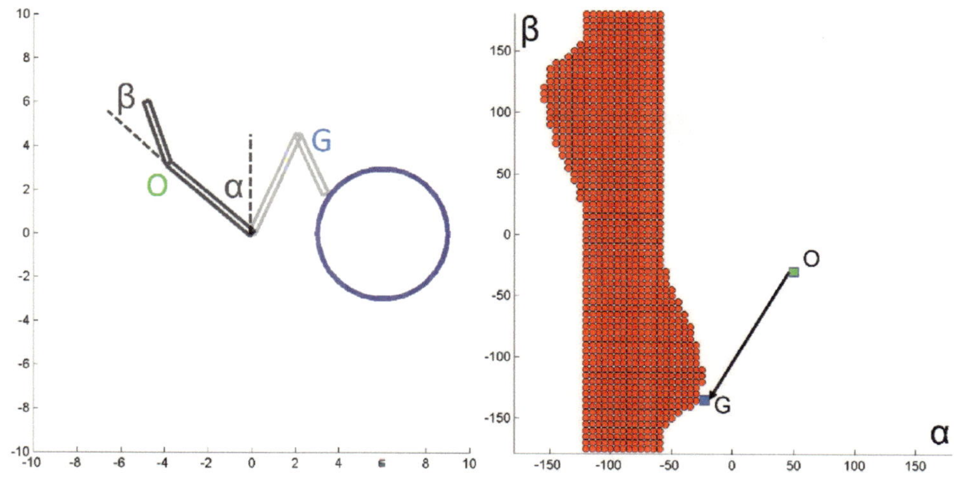

Figure 1.
C-map of a planar robot defined by 2 links and 2 rotational joints.

2.1 Obstacle definition

Objects have many possible shapes in real environment, but most of them can be accurately represented by the sum of a finite number of convex shapes. For sake of simplicity, only 2D objects are taken into account in this chapter, but similar statements can be derived for 3D objects. Since 3D objects have one more dimension, the computational costs of the algorithms should generally increase by one order of magnitude. In some cases, this computational cost variation makes algorithms feasible for 2D object scenarios and not feasible for 3D object scenarios.

Robots have basically two ways to include obstacles into their path planification:

- Detecting obstacles by themselves.
- Having another system which constantly provides obstacles to them.

Object detection is a very broad field and many sections of it are still opened. As E. Arnold et Al reports in their paper [7], there are a lot of different techniques to detecting objects according to sensor types, sensor configurations and the operative scenarios. Due to this reason, this chapter is not going to deal with object detection. All the objects in the scene are assumed to be known and provided by another system to the path planner.

As it was mentioned before, this chapter deals with 2D objects to ensure 2D visualization. 2D objects can be convex or non-convex. As it is shown in **Figure 2**, convex objects do not have any internal angle larger then 180°. Non-convex objects, on the other hand, have at least one of them. Non-polygonal objects are convex by default if they are regular shapes (circles, ellipses, etc.…). If objects are defined by a concatenation of various curves, a possible solution to establish whether the object belongs to convex or non-convex shapes is rearranging the edges of the object by a set of liner segments. According to the resolution of this edge substitution, the computational cost of convex categorization changes. Higher resolution leads to high precision, but high computational cost.

This convex categorization is fundamental because computing convex object collision is very easy and not computationally expensive due to computational

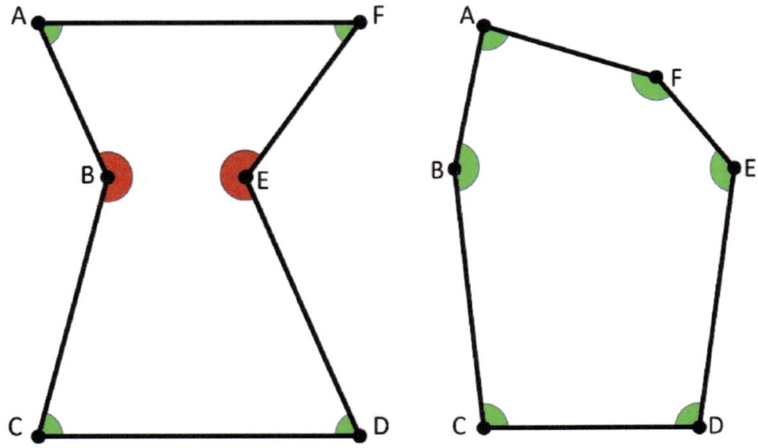

Figure 2.
Non-convex and convex 2D objects.

geometry algorithms [8]. So, it is better to "convert" non-convex shapes to convex shapes. The most used techniques to perform this transformation are:

- Convex hull algorithms (e.g., Graham's scan)
- Subdivision algorithm
- Triangle meshing algorithm

The *Convex hull algorithms* are able to find the smallest convex hull which includes all points of a dataset. As it is described in [8], there are many algorithms which are able to tackle this problem. The *Graham's scan* is probably one of the most famous. Since the corners of a shape can be seen as the points of a dataset (hull), the Graham's scan would be able to wrap all of them into a unique hull erasing all critical corners. **Figure 3** shows that the non-convex shape (ABCDEF) would be converted into a convex one (ACDF) by erasing corners B and E.

The *Convex hull algorithms* are techniques quicker than *Subdivision* if the number of critical corners is quite large, but they significantly reduce the accuracy of the object definition. Practically, those techniques cut out all critical corners and enlarge the area occupied by the original shape. The Graham's scan works perfectly on obstacles defined by a point cloud which, generally, derives from raw data. Since this chapter deals with known objects, there is no reason to take this technique into account.

The *Subdivision algorithm* is a technique which aims to redefine the original non-convex shape into the sum of convex sub-shapes. It does not compromise the original occupied area, because the algorithm tries split critical corners by dividing the original shape into sub-shapes. At each iteration (subdivision), the algorithm has to check whether the new shapes are still non-convex. If yes, the algorithm has to continue the operations. The main drawback of this algorithm is that the final number of sub-shapes is never known a priori. The only known thing is that the larger number of sub-shapes is p-2, where p is the number of corners. In the worst-case scenario, the algorithm has to subdivide the original shape into triangles which are always convex.

The *Subdivision algorithm* is theoretically the best technique because it splits just the critical corners until all shapes are convex, At the end of the process,

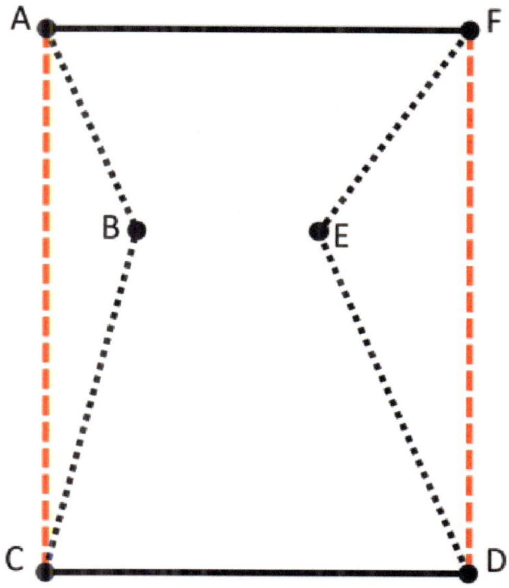

Figure 3.
Non-convex shape converted into a convex one by convex hull algorithms.

the number of sub-shapes is the smallest to ensure the overall convexity, but the algorithm has to check at every iteration which sub-shapes are still non-convex. The process might cost quite a lot according to the original object shape. **Figure 4** reports an example of shape *Subdivision algorithm*.

The *Triangle meshing algorithm* might be seen as special case of the *Subdivision algorithm*. Practically, it subdivides the original shapes until all sub-shapes are triangles like the worst-case scenario of the *Subdivision* technique, but it does not check sub-shapes convexity. Triangles are convex by definition.

The *Tringle meshing algorithm* is quick in term of subdivision if the number of critical corners is quite large because all sub-shapes created by the algorithm are

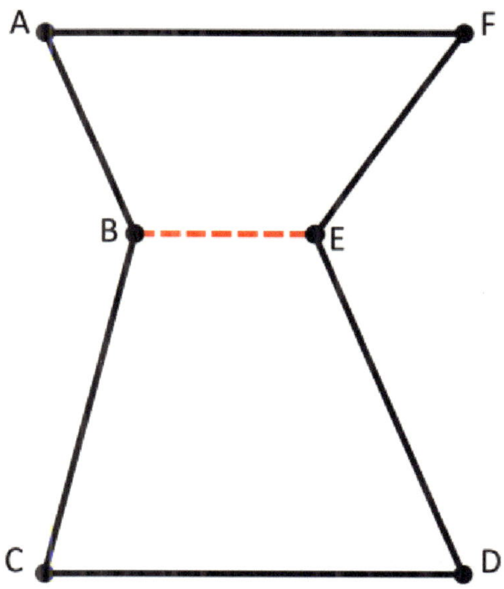

Figure 4.
Non-convex shape converted into a convex one by subdivision algorithm.

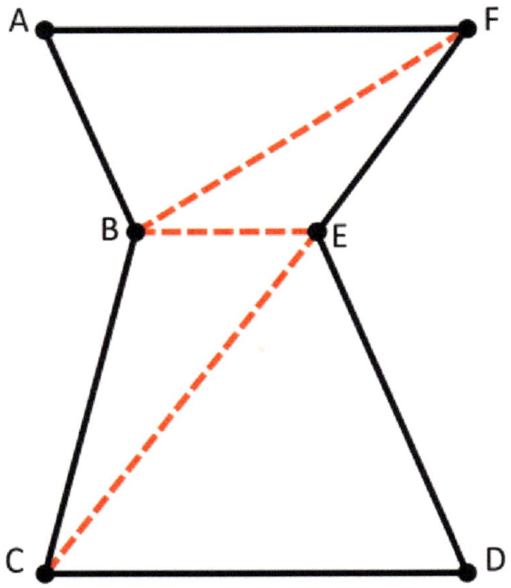

Figure 5.
Non-convex shape converted into a convex one by triangle meshing algorithm.

triangles. Since triangles are convex by default, there no reason to check whether sub-shapes are convex or not. The major advantage is that the number of iterations is related to the number of corners of the original object. In particular the number of sub-shapes is p-2, where p is the number of corners. **Figure 5** reports an example of this algorithm.

The major drawback of the *Tringle meshing algorithm* is related to the large number of final sub-shapes compared to the *Subdivision* technique. This drawback definitely affects object collision process, because having more objects means performing more collision checks during the c-map creation.

Since this chapter deals with known non-deformable objects, using the *Tringle meshing* technique is quite inefficient. The path planner converts non-convex shapes into convex just one time at the beginning of the whole process. Afterward, it will extensively check object collisions. Having the lowest amount of objects in the scene is crucial to reduce computational cost. To conclude, The *Subdivision algorithm* appears to be the most efficient.

2.2 C-space computation

According to the number of degrees of freedom of the robot, the type of degrees of freedom of the robot, the number of objects in the scene, and the shape type of those objects (circle, polygonal, etc....), computing the "*collision free c-space*" changes significantly. Moreover, the computational cost will increase according to the number of objects in the scene and the number of degrees of freedom. In theory, there is an equation which describes the c-space, but defining it is often very complex.

Let us consider a trivial example: a robot defined by a single link and a single rotational joint. The rotational joint is located to one of its ends and the link can only rotate around one of it end; no translation is involved. Moreover, there just one object in the scene defined by a circle (**Figure 6**).

Computing the c-map of this trivial example is immediate: there is just one degree of freedom represented by the angle α of the rotational joint. Since the

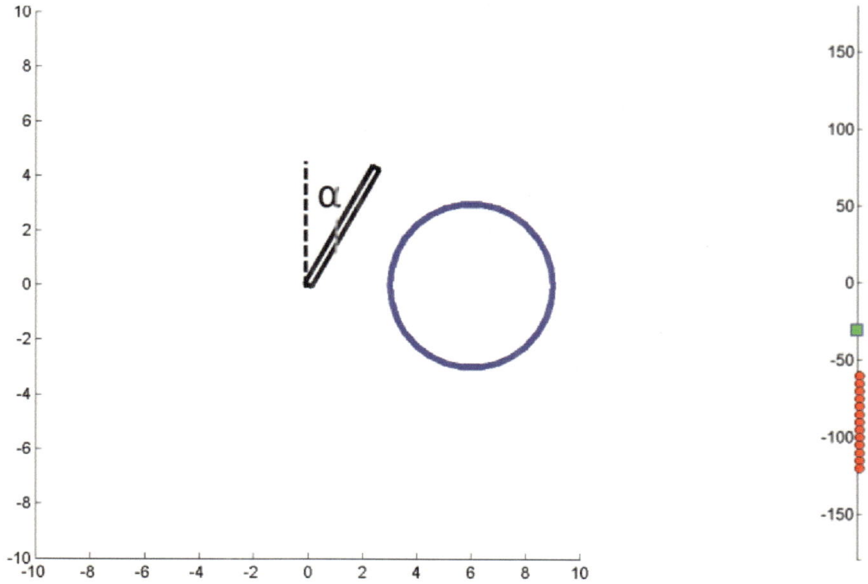

Figure 6.
C-map of a planar robot defined by 1 link and 1 rotational joint.

obstacle is just one, it is sufficient to calculate for which values the link touches the obstacle without crossing its edges to know which is the valid angle range. The c-map of **Figure 6** shows that the *collision free c-map* is basically a bar going from −180° to −120° and from −60° to +180°. The robot collides with the obstacle while α is included in [−120°; −60°] (**Figure 7**).

The computational cost required to compute this c-map is almost zero.

Another trivial example is related to robots able to translate on a 2D map. In the case of a robot represented by a single point, the *collision free c-map* corresponds exactly to the areas not occupied by the obstacles. If robots are defined by other shapes (circles, polygons, etc.), calculating the c-map might looks like trickier than the previous example, but there is a practical way to do it. The algorithm has to make the robot slide around every obstacle to the define the boundaries of the

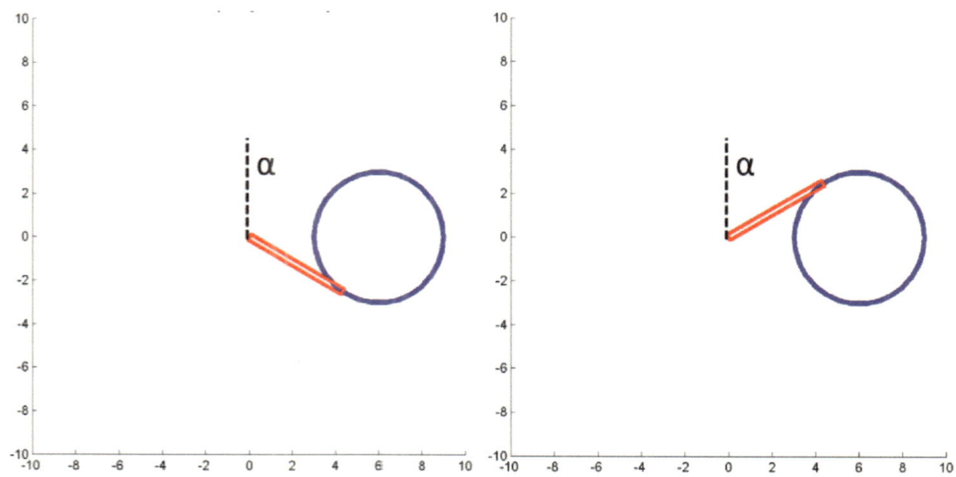

Figure 7.
Planar robot position for α equal to −120° and −60°.

collision free c-map. Afterwards, the algorithm has to fill those area to reconstruct the full map. **Figure 8** shows an example of a polygonal robot able to translate along both X and Y axes in [−8; +8]. In the scene there are two obstacles: a circle and a polygon. The robot cannot rotate around the z-axis and its location is referred to the position of the yellow point placed on one of its edges. The light-blue square represents in **Figure 8** is the area where the yellow point can move.

Let us take into account one of the most popular c-map examples: a two link planar robot having two degrees of freedom represented by the angles of the rotational joint connecting the first link to the ground and the rotational joint connecting the first link to the second one. The simplest scanario includes just one obstacle represented by a circle (**Figure 9**).

Due to the complexity of defining a single equation which describes accurately the *collision free c-map*, the algorithm had to simulate all possible configurations of the robot and check if it would collide with the obstacle. Obviously, the number of possible robot configurations are infinite because each degree of freedom is defined on the interval (−180°; +180] ∊ ℝ. In order to solve this issue, the algorithm had

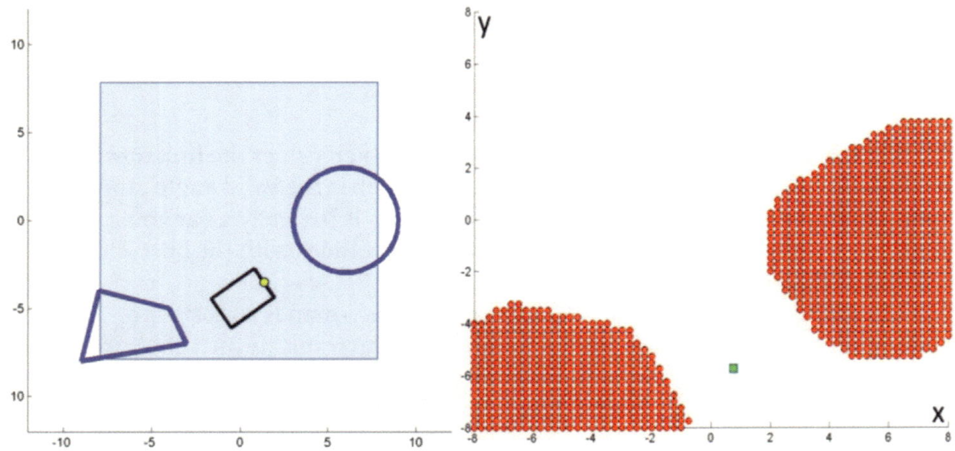

Figure 8.
C-map of a polygonal robot able to translate on a 2D map.

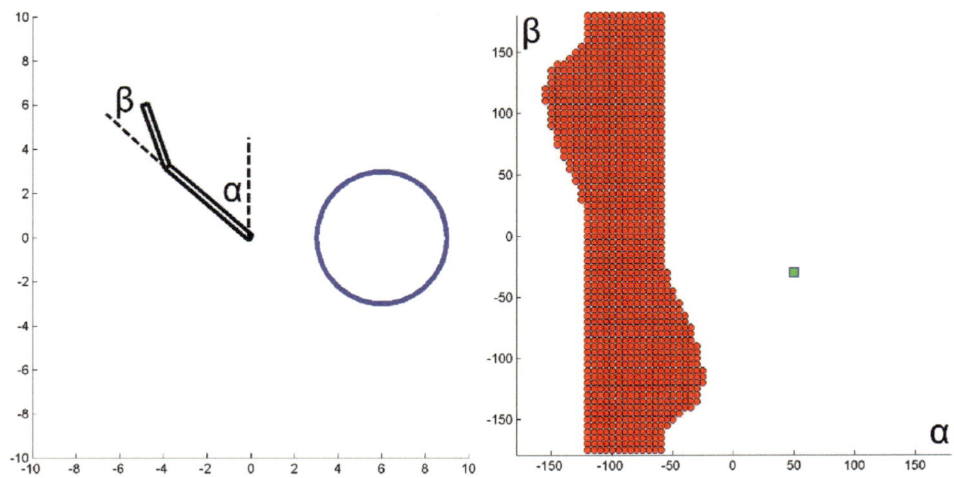

Figure 9.
C-map of a planar robot defined by 2 links and 1 circular obstacle.

to discretize that interval. The c-map represented in **Figure 9** has been computed by discretizing each degree of freedom by 5°. At this resolution, the whole process required 5148 iterations and it took 8 seconds on the test machine. On average, just 1 second have been used to evaluate object collision. The other 7 seconds have been used for reading and writing memory cells.

Before continuing with the chapter, is better to highlight the fact that every experiment has been performed on the same machine using a mono-thread fashion algorithm in order to avoid the possibility of using too powerful computers. Some algorithms, such as the c-map simulation search, have each iteration totally unrelated to the previous ones. This fact gives scientists the possibility to project a system which uses as many cores as the number of expected iterations to return the algorithm result immediately. Unfortunately, this approach might work in theory, but it is practically impossible for most of the cases because it would force the robot to be linked to a supercomputer which cannot be contained inside of the robot. Let us imagine that having an external facility containing this large computer would be feasible and that the algorithm should calculate the *collision free c-map* of a robot shake which has 9 links and each link is connected to the following one by a 3 axes rotational joint. Moreover, since the robot has to face a very narrow scenario, each degree of freedom has to be discretized by at least 1000 samples. The algorithm should iterate $\mathbf{1000^{3*9}}$ which is approximately the number of atoms of the universe.

In order to have a better understanding of the computational resources required to compute the *collision free c-map*, let us take into account the same two link planar robot, but with just one polygonal obstacle in the scene (**Figure 10**). Finding whether the robot collides with a polygonal obstacle is way more expensive as described in book [9, 10]. The number of edges of the obstacle covers an important role, but in this chapter every polygonal obstacle is assumed to be defined by no more than 5 edges for sake of simplicity.

Creating the *collision free c-map* of this scenario required the same amount of iteration as the scenario represented in **Figure 9**, but the whole process lasted 46 seconds. Similarly to the previous example, 39 seconds have been used to evaluate object collision and 7 to write and read memory cells. Computing c-map on a 5-edge polygon obstacle was 39 times longer than computing it on a circular obstalce.

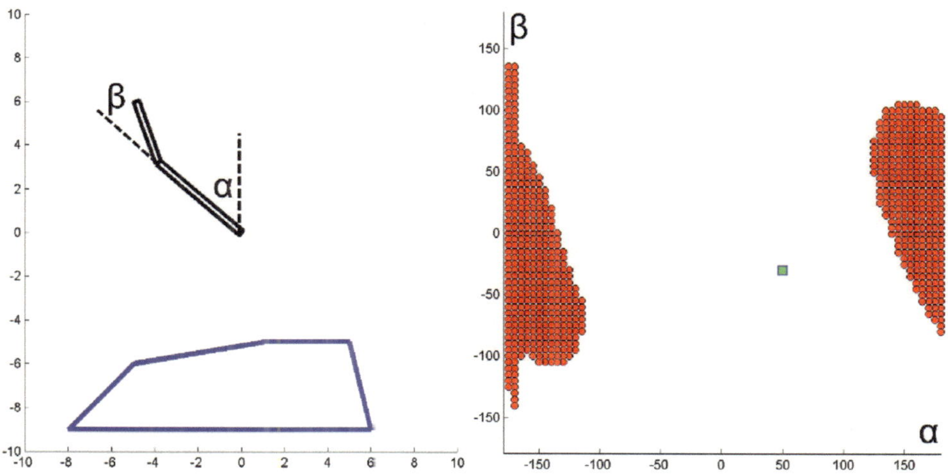

Figure 10.
C-map of a planar robot defined by 2 links and 1 polygonal obstacle.

Adding one circular object in the scene means increasing the computational cost by 1 while adding a polygonal obstacle means increasing the computational cost by 39.

In order to prove that linearly increasing the number of degrees of freedom will exponentially increase the computational cost, a bunch of experiments have been run on three link planar robot. The number of iterations required to find all possible robot configurations with a resolution of 5° is 373248. Calculating the *collision free c-map* of a scenario including one polygonal and one circular obstacle lasted almost 5000 seconds on the machine used to run all experiments reported in this chapter.

2.3 C-space usage

As soon as the c-map has been computed, another piece of the algorithm takes the lead and tries to find the path leading the robot from its initial configuration to the goal one. Sometimes, the path is a single straight line connecting the original point to the goal point on the c-map, but most of the times, the path is defined by a set of segments (**Figure 11**); each point connecting a segment to the following one is called way-point. Unfortunately, not all scenarios have a solution: some c-maps have more than just one obstacle free sector. If the robot is located into one of them, it cannot jump into the other one (**Figure 11**).

As it was mentioned in Section 2.2, the algorithm in charge of computing the c-map had to perform the colliding check of all representative configurations of the robot which means that the algorithm had create a grid of cells defined by two parameters: their position into the map (robot configuration) and a boolean value which states whether the robot collides with an obstacle or not.

Mathematically, the algorithm converts a $\mathbb{R}^n$ space into a $\prod_n d_i$ space, where n is the number of degrees of freedom and d_i is the number of samples for the ith degree of freedom. The algorithm had basically clustered the infinite states of a system into the most representative ones.

According to S. M. LaValle [2], the are many ways to compute a path starting from the c-map. Let us take into account, as an example, the potential field path planner. Briefly, the potential field path planner associates to each cell of the c-map to another value which comes from the sum of the attractive potential field and the repulsive potential field. The first one (a) guides the robot configuration to the goal one, while the second one (b) pushes the robot configuration far from obstacles (**Figure 12**).

Afterward, starting from the initial configuration cell, the algorithm has to check which cell among the surrounding ones has the lower potential value and takes that as the last explored cell. Then, the algorithm has to repeat the search until

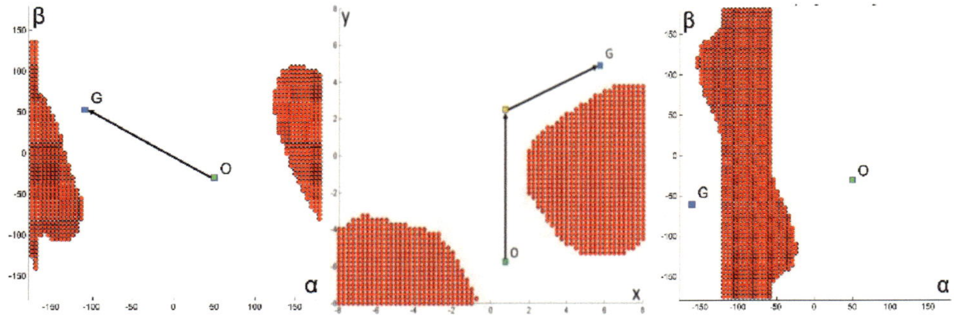

Figure 11.
Typical 2D path solutions.

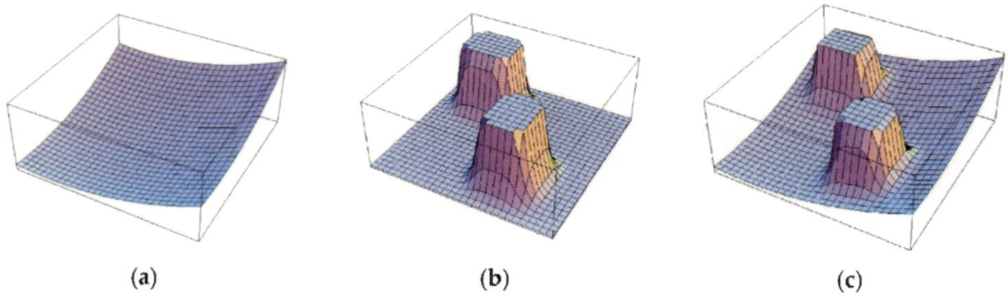

Figure 12.
Graphical visualization of a generic 2D potential field map having two obstacles in the scene (attractive field + repulsive filed) [11]. (a) shows the component of the potential field that attracts the robot towards the goal configuration; whereas (b) is the component that pushes the robot away from the configurations in which it would collide with the obstacles; (c) shows the final combined potential field.

the goal cell has been reached or it gets stuck into a local minima. This approach is no longer the state-of-the-art because it has many drawbacks which are not reported in this chapter. However, it has been used as an example because it is very simple and gives the perfect picture of performing a path search on a cell map.

As it was shown in **Figure 11**, creating the path by analyzing cell by cell might be very inefficient. Sometimes, there is even a straight line connecting the goal configuration and the initial one, but the algorithm has to waste time performing the whole c-map computation anyway. A possible strategy to overcome this issue is using a heuristic method.

3. Heuristic methods

Heuristic methods become very popular because they are able to find a good solution (non-optimal) in a very short time. They are specifically popular to for solving scheduling problems. Heuristic methods are not analyzing the problem step by step like most of the deterministic algorithms; they are starting directly from a potential solution. Then, they adjust this solution iteratively in order to reduce the outcome of a cost function. Heuristic algorithms look for alternative solution randomly or pseudo-randomly at every iteration until a certain time or number of iterations has been reached. In [12] a variety of heuristic methods are described. Scheduling problems find a perfect match for these algorithms because sorting randomly all items the algorithm have to schedule is already a solution. The algorithm is not supposed to create or erase some items. Path planning algorithms, instead, do have to create a set of waypoints connecting the initial configuration to the goal one on the c-map. Way-points are a set of scheduled configurations that the robot has to cross sequentially in order to reach the goal. In path planning algorithms way-points are the items to be scheduled.

Nowadays, heuristic algorithms able to create or erase items from the original schedule pool are included in the family of genetic algorithm [13, 14]. Within each generation, the number of items of the schedule is constant. From one generation to another one the number of items might change.

Coming back to the potential field planner, it practically tries to connect the initial configuration and the goal configuration by selecting the less expensive cell in term of potential field at every iteration. Each cell is a sort of waypoint that the robot has to cross. The outcome of the cost function related to potential field planners is the sum of the potential costs of all cells the algorithm plans to cross. The potential field planner deterministically finds the path by analyzing the map

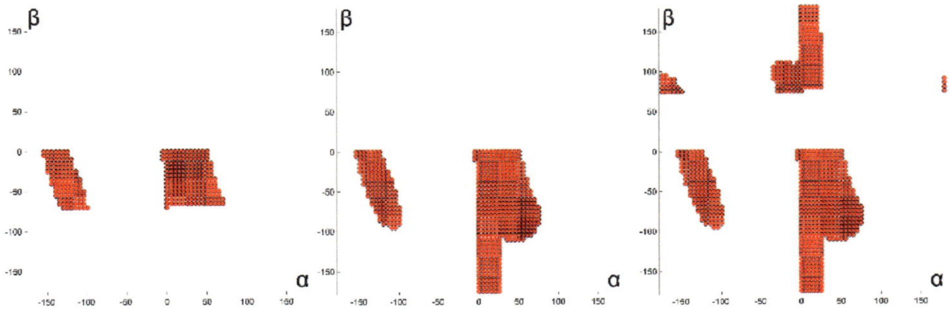

Figure 13.
C-map computation progresses (20%, 50%, 80%).

step by step. An alternative way to find a feasible solution is drawing a straight line connecting the goal point to initial one at first. Afterwards, if the path intersects an obstacle of the c-map, the algorithm should introduce a waypoint and making it slide on the map randomly for some iterations. If a feasible solution has not been found, the algorithm should introduce another waypoint and so on. The reader of this chapter might argue that this process might get stuck if there are no solutions because there is no way for the algorithm to understand it deterministically. (S) He would be right, but this is the drawback of using heuristic algorithms. They are very quick to find solutions if they exist, but the algorithm's target is not finding the optimal solution. However, if the algorithm is well designed heuristic algorithms are one of the most powerful weapons to tackle path planning problems quickly.

Unfortunately, the heuristic algorithms described so far are performing the path search on the c-map. It means that the c-map has to be computed in advance anyway. Apparently, using heuristic methods speeds up the path search, but do not solve the time issue raised up by the *collision free c-map* computation appointed in Section 2.5. However, looking closer to the process of creating the c-map the reason of such a time-consuming process is obvious: the c-map computation algorithm is deterministically discretizing all possible robot configurations. At the end of the iterations the statistical distribution of configuration samples is perfectly flat, but it is unbalanced for the whole process. **Figure 13** shows the progresses of a c-map computation (20%, 50%, and 80%).

The accuracy of the map is very high where the c-map has been computed, while it is null where the map has still to be processed. A feasible way to keep the accuracy homogenous over the whole map during the entire computation process is computing cells randomly on the map. Due to the "Monte Carlo's simulation", picking points randomly on a dataset converges quite quickly to the statistical distribution of that dataset. In this case, the dataset are the cells of the map and their statistical distribution is flat because cells are equally distant from each other.

At this point, a quite simple question might come out: "Is analyzing all cells mandatory in order to find a feasible path?". The answer, of course, is "no". This process of configuration space discretization is required to create a full c-map, but even rarefied c-map can lead to feasible and acceptable paths. The family of sampling based planners is the result of this idea.

4. Sampling based planners

Sampling based planners finds their strength into reducing dramatically the number of robot configurations taken into account during the c-map construction process. Instead of creating the full c-map, few points are randomly selected into the map in order to reduce the number of robot configuration taken into account without loosing the statistical distribution of the map. The most common algorithms belonging to this family are the probabilistic Roadmap (PRM) [3, 4] and the Rapidly-exploring Random Tree (RRT) [5, 6]. Both algorithms are throwing into the scene random points one by one and tries to connect them to the closer one in order to build a graph (PRM case) or a tree (RRT case). A connection is considered valid if the line connecting a point to another one is not crossing an obstacle. Practically, the algorithm is locally computing a c-map while it tries to connect two points. At first, it might look not so efficient because the algorithm seems to replicate the classical c-map construction with the drawback of having just a graph or a tree and not the full map. However, as soon as the number of degrees of freedom overpasses 2, the computational cost difference between building the full c-map and using a sampling based planner the significantly increases.

The first is optimal for multi agent planning algorithms because graphs do not have a starting node. The initial configuration of the robot can be located anywhere on the map and linked one node of the graph. Similarly, the goal configuration can be linked to a node of the graph. Afterwards, a graph search algorithm (Dijkstra's algorithm [15], A* [16], etc....) will be in charge of finding the best path on request. Exploring trees, instead, are based on node hierarchy. Every node is connected the others using a father-child fashion hierarchy. Every node must have a father except for the root node. The root node is the one corresponding to the initial robot configuration. This means that exploring the tree is extremely very fast and simply but, the algorithm has to compute the tree if the initial robot configuration changes.

5. Conclusions

Nowadays, heuristic and meta-heuristic methods are widely used because they are incredibly efficient in term of computational cost. Moreover, if they are well designed, they are able avoid local minima which are far from the global minima. Philosophically, heuristic methods are the family of problem-solving algorithms closer to human thinking. Humans uses to solve problems by attempting it practically, simulating it in their mind, or doing both at the same time. However, humans require much more time to perform this search than computers. This is one of the main reasons why drug development is so quick today. Scientists do not practically mix compounds all the time; most of their job is simulating chemical reactions.

Computing c-maps deterministically is very expensive or even impossible for a large variety of scenarios. Heuristic methods appear to be a solution because they are very quick and allows scientists to have at least a non-optimal feasible path. So, the sampling based planner family comes from the necessity of unifying c-map computation and heuristic methods into a new path planning technique.

Author details

Emanuele Sansebastiano and Angel P. del Pobil*
Robotic Intelligence Lab, Jaume I University, Castellón de la Plana, Spain

*Address all correspondence to: pobil@uji.es

References

[1] M. Elbanhawi and M. Simic, "Sampling-based robot motion planning: A review," in IEEE Access, vol. 2, pp. 56-77, 2014, DOI:10.1109/ACCESS.2014.2302442.

[2] Cambridge University Press. "Planning Algorithms", by S. M. LaValle. Available at http://planning.cs.uiuc.edu/; 2006.

[3] L. E. Kavraki, P. Svestka, J. Latombe and M. H. Overmars, "Probabilistic roadmaps for path planning in high-dimensional configuration spaces," in IEEE Transactions on Robotics and Automation, vol. 12, no. 4, pp. 566-580, Aug. 1996, DOI:10.1109/70.508439.

[4] Yan, F., Liu, YS. and Xiao, JZ. Path planning in complex 3D environments using a probabilistic roadmap method. Int. J. Autom. Comput. 10, 525-533 (2013). DOI:10.1007/s11633-013-0750-9

[5] LaValle, S. "Rapidly-exploring random trees: A new tool for path planning." The Annual Research Report. 1998

[6] Rodriguez, Xinyu Tang, Jyh-Ming Lien and N. M. Amato, "An obstacle-based rapidly-exploring random tree," Proceedings 2006 IEEE International Conference on Robotics and Automation, 2006. ICRA 2006., 2006, pp. 895-900, DOI:10.1109/ROBOT.2006.1641823.

[7] E. Arnold, O. Y. Al-Jarrah, M. Dianati, S. Fallah, D. Oxtoby and A. Mouzakitis, "A survey on 3D object detection methods for autonomous driving applications," in IEEE Transactions on Intelligent Transportation Systems, vol. 20, no. 10, pp. 3782-3795, Oct. 2019, DOI:10.1109/TITS.2019.2892405.

[8] de Berg, Mark and Cheong, Otfried and van Kreveld, Marc and Overmars, Mark; 2010. "Computational geometry: Algorithms and applications (3nd edn.)", Springer Publishing Company. ISBN 3642096816 DOI:10.5555/1951877

[9] Bayer, Valentina. "Survey of Algorithms for the Convex Hull Problem," Technical report, Oregon State University; 1999.

[10] Gamby, A.N.; Katajainen, J. Convex-Hull algorithms: Implementation, testing, and experimentation. Algorithms 2018, 11, 195. DOI:10.3390/a11120195

[11] Jeon, G.-Y.; Jung, J.-W. Water sink model for robot motion planning. Sensors 2019, 19, 1269. DOI:10.3390/s19061269

[12] Silver, E. "An overview of heuristic solution methods". J Oper Res Soc 55, 936-956, 2004. DOI:10.1057/palgrave.jors.2601758

[13] Beasley, David, Bull, David R. and Martin, Ralph Robert. "An overview of genetic algorithms: Part 1, fundamentals". University Computing 15 (2), pp. 56-69. 1993

[14] Lingaraj, Haldurai. "A study on genetic algorithm and its applications". International journal of computer sciences and Engineering. 4. 139-143. 2016

[15] Dijkstra, E.W. "A Note on Two Problems in Connexion with Graphs." Numerische Mathematik 1 (1959): 269-271. <http://eudml.org/doc/131436>.

[16] P. E. Hart, N. J. Nilsson and B. Raphael, "A formal basis for the heuristic determination of minimum cost paths," in IEEE Transactions on Systems Science and Cybernetics, vol. 4, no. 2, pp. 100-107, July 1968, DOI:10.1109/TSSC.1968.300136.

Chapter 4

A Survey on Recent Trends of PIO and Its Variants Applied for Motion Planning of Dynamic Agents

Muhammad Shafiq, Zain Anwar Ali and Eman H. Alkhammash

Abstract

Pigeon Inspired Optimization (PIO) algorithm is gaining popularity since its development due to faster convergence ability with great efficiencies when compared with other bio-inspired algorithms. The navigation capability of homing pigeons has been precisely used in Pigeon Inspired Optimization algorithm and continuous advancement in existing algorithms is making it more suitable for complex optimization problems in various fields. The main theme of this survey paper is to introduce the basics of PIO along with technical advancements of PIO for the motion planning techniques of dynamic agents. The survey also comprises of findings and limitations of proposed work since its development to help the research scholar around the world for particular algorithm selection especially for motion planning. This survey might be extended up to application based in order to understand the importance of algorithm in future studies.

Keywords: Pigeon Inspired Optimization, Dynamic Agents, Optimization, Bio-Inspired Computation, Motion Planning Techniques

1. Introduction

The searching ability of homing pigeon is unmatched with other birds as it can be more accurate to achieve the destination despite long distance traveling [1]. Therefore homing pigeons have been widely used in 18th century to send and receive mails from far distances with minimal errors. As the telecommunication became popular for sending and receiving mails, the use of Pigeons almost vanished. With the advancement of technology, complex systems seek more accurate and stable algorithm to sort the convergence and stability issues.

The homing behavior of pigeons uses global searching ability to find the target with the help of natural navigation parameters i.e. Sun and Earth's magnetic field [2]. Initial studies on pigeons suggest that the pigeon can find the difficult destinations in most easy way when compared to other similar spices [3]. According to studies, the species appears to have a mechanism in which signals from magnetite particles are conveyed by the trigeminal nerve from the nose to the brain [4]. The capacity of pigeons to perceive varied magnetic fields was investigated, and it was discovered that the pigeons' amazing homing skills are nearly entirely reliant on small magnetic particles in their bills. Pigeons have iron crystals in their bills, which can give them a sense of direction [5, 6]. The flying direction of the moving bird is tuned by relative orientation mapped by two basic operators [7, 8]. **Figure 1** shows the basic approach used by the pigeons to map the route to destination and coming back to home [9].

Based on the searching ability for global search and route planning in pigeons encourage the researcher to introduce a novel optimization algorithm namely "Pigeon Inspired Optimization (PIO) algorithm" in 2014 for optimal solutions [10]. Further improvements in existing algorithm have been made time to time to concur variety of optimization problems including aerial field.

The unwanted uncertainties and complexities of various agents formed in a group still challenging for many researchers. To improve these hurdles proper motion planning of agents required that can reduce the convergence time and enhance stability of the system. The basic PIO has many improvements in its structure as well as combined with related algorithms in order to improve performance and stability of complex systems. This study sum up the motion planning techniques based on PIO and its variants of many agents including unmanned aerial vehicles (UAV's) with the help of findings and limitations. The works is based on open access PIO papers and its variants found on scholar portal of Google till May 2021.

Further layout of the chapter is as follows: Section 2 present novel idea in optimization problems. Section 3 discusses the mechanism and principle of PIO. Section 4 reviews the basic PIO and its variants applied on dynamic agents. Section 5 explains the conclusion and future work.

2. State of art

The state-of-the-art and intelligent optimizer has been introduce by Duan and Qiao and termed as Pigeon Inspired Optimization (PIO) Algorithm. This algorithm is based on homing behavior of pigeons that used simplified concept of route following either to detect target or coming back to home. The pigeons use earth

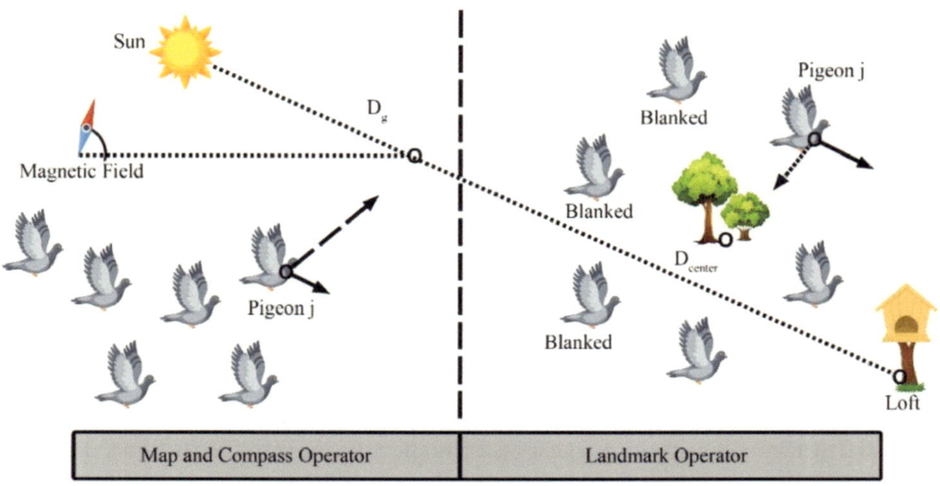

Figure 1.
Pigeon's homing behavior mechanism [9].

magnetic field, sun and landmarks for their complete journey as navigation tools. The basic PIO algorithm uses conventional mathematical expressions for "Map and Compass operator" and "Landmark operator" to produce navigation system.

Pigeons use magnetic based receiver to configure the map in their brains to perceive the earth field. To adjust the direction of their route they prefer the sun's elevation when available. They rely less on the sun and magnetic particles as they fly to their target. When the pigeons are getting close to their goal, they will rely on nearby landmarks. If they recognize landmarks then they can move fast and use direct route same as previous one. Now if any pigeon does not recognize landmark then they find one who is familiar with landmark and started following them.

3. Preliminaries of pigeon inspired optimization

3.1 Mechanism

In nature, homing pigeons use very simple navigation mechanism to find their homes. This mechanism is based on sunlight and pigeon's own shadow to trace out suitable route to destination. This mechanism is being very famous among active researcher around the globe. Moreover, this mechanism does not only depend upon the sun therefore other factor must be included to avoid errors in overcast condition or when the sun is not available.

Navigation mechanism of homing pigeon disturbed when the sun is hidden and unable to provide proper navigation the earth magnetic field becomes another navigation tool in order to maintain her flight. Since 2014, when this mechanism was first introduced by DUAN, researcher in the field validates that the magnetic field theory is being perfect tool for navigation.

3.2 Principle optimization

In Pigeon Inspired Optimization, a natural mechanism exists through which a pigeon can trace the path from initial point to the target. After years of studies it can be found that the pigeons are the most suitable bird for target detection, path planning and faster convergence related issues in optimization based problems [9, 10].

To obtain mathematical expression of PIO algorithm there are two separate operators i.e. the map & compass operator and the landmark operator; these operators describe the navigational effects of the sun and Earth's magnetic field, as well as that of familiar landmarks, respectively.

Suppose there is M pigeons are moving in the air space forming search space. When map and compass operator contain $Mc \leq Mc^1_{max}$, iteration for every pigeon's navigation j providing Mc^1_{max} is the maximum iteration and D_j^{Mc+1} is the position of pigeon j at iteration $Mc + 1$ is updated by

$$\begin{cases} V_j^{Mc+1} = e^{-R\,(Mc+1)}.V_j^{Mc} + rand.\left(D_g - D_j^{Mc}\right), \\ D_j^{Mc+1} = D_j^{Mc} + V_j^{Mc+1}, \end{cases} \tag{1}$$

where V_j^{Mc} and V_j^{Mc+1} represent j pigeon's velocities at iteration Mc and Mc +1, respectively, R shows map and compass factor, rand variable used for random number [0,1], Dg for global best position, and D_j^{Mc} is the pigeon's position at iteration Mc.

The navigation system of pigeon is presented by landmark operator when $Mc^1_{max} \leq Mc \leq Mc_{max}$ Where Mc_{max} for maximum iteration of PIO and fulfills the condition $Mc_{max} \leq \log_2(N) + Mc^1_{max}$. The position function D_j^{Mc+1} is expressed as in the following equation:

$$\begin{cases} M = \lceil M/2 \rceil, \\ D^{Mc}_{center} = \dfrac{\sum_{i=1}^{M} D_j^{Mc}.\mu\left(D_j^{Mc}\right)}{\sum_{i=1}^{N} \mu\left(D_j^{Mc}\right)} \\ D_j^{Mc+1} = D_j^{Mc} + rand.\left(D^{Mc}_{center} - D_j^{Mc}\right) \end{cases} \quad (2)$$

where $\lceil \cdot \rceil$ is used for ceiling function. D^{Mc}_{center}, is the average weighted landmark positions at iteration Mc. The weight $\mu\left(D_j^{Mc}\right)$ is calculated by:

$$\mu\left(D_j^{Mc}\right) = \begin{cases} f\left(D_j^{Mc}\right), \text{for maximization}, \\ \dfrac{1}{f\left(D_j^{Mc}\right) + \varepsilon}, \text{for minimization}, \end{cases} \quad (3)$$

wheref $\left(D_j^{Mc}\right)$ shows the cost function pigeon j at iteration Mc with any nonzero constant.

4. Pigeon inspired optimization and its variants

In the world of artificial intelligence, intelligent algorithms are needed to be changed time to time in order to maintain precise optimization and complex problem identifications. A number of optimization algorithms have been used to counter these problems such as Ant Colony Optimization (ACO), Genetic Algorithm (GA), Artificial Bee Colony (ABC), Particle Swarm Optimization (PSO) and Pigeon Inspired Optimization (PIO) etc. have been widely used in many optimization problems. PIO is the state of the art optimization algorithm that was initially proposed for aerial robot path planning problems. Due to its simplicity and optimizing ability, PIO has been combined with other algorithms to avoid trapping into local optima as well as faster response. Furthermore, modifications in basic PIO algorithm based on structure, operation and application has been gathered in **Table 1** to review for motion planning of multiple agents. Year wise distribution of PIO variants are as follows.

In 2014, Duan and Qiao [10] introduced a novel optimization process termed as Pigeon Inspired Optimization (PIO) algorithm for path planning of aerial system. This novel algorithm comprises of multiple self-governing operators: map and compass operator for magnetic field effect of earth and landmark operator for remembering the route with the help natural behavior of homing pigeons. Zhang and Duan [11] proposed a novel Predator–prey pigeon-inspired optimization (PPPIO) for 3-D path planning problem solution of unmanned aerial vehicles (UAVs). Zhang and Duan [12] again proposed improved PIO: PPPIO for 3-D path planning of Uninhabited Combat Aerial Vehicle. Li and Duan [13] achieved low altitude target detection for UAVs with the help of hybrid algorithm of Simulated

S. No	Ref	Title	technique used	applied on	compared with	key findings	limitations
1.	[10]	"Pigeon-inspired optimization: a new swarm intelligence optimizer for air robot path planning"	A new algorithm based on Pigeon's natural behavior namely Pigeon Inspired Optimization Algorithm (PIO)	UAVs	DE	• Faster convergence and optimize global search. • Generate smooth optimal path planning.	• Not for complex path planning problems. • PIO algorithm is not valid for confined ideal solution.
2.	[11]	"Predator–Prey Pigeon-Inspired Optimization for UAV Three-Dimensional Path Planning"	Pigeon Inspired Optimization with Predator Prey(PPPIO)	UAVs	PIO PSO	• Improved population diversity. • Increased the convergence speed.	• Not valid for high number of iterations Variations occurred after achieving stability.
3.	[12]	"Three-Dimensional Path Planning for Uninhabited Combat Aerial Vehicle Based on Predator–Prey Pigeon-Inspired Optimization in Dynamic Environment"	Prey–Predator PIO algorithm (PPPIO)	Uninhabited Combat Aerial Vehicle	PIO PSO DE	• Best cost function achieved of the system • Best efficiency and convergence speed achieved	• Insufficient for fixed-wing aircrafts • Application missing
4.	[13]	"Target detection approach for UAVs via improved Pigeon-inspired Optimization and Edge Potential Function"	Simulated Annealing based Pigeon Inspired Optimization algorithm with Edge Potential Function (SAPIO-EPF)	UAVs	GA PSO ABC PIO	• Enhance convergence speed. • Robust target detection of UAV at low altitude.	• Higher computational time due to image size. • Slower convergence speed than basic PIO
5.	[14]	"Pigeon-Inspired Optimization Approach to Multiple UAVs Formation Reconfiguration Controller Design"	Pigeon Inspired Optimization with Control Parameterization and Time Discretization (PIO-CPTD)	UAVs	PSO	• Achieved same height maneuvering • Capability to discover smaller value than PSO.	• Need extra calculations for discretization. • Achieved local minima after 600 iterations
6.	[15]	"Multiple UAVs Mission Assignment Based on Modified Pigeon-Inspired Optimization Algorithm"	Modified Pigeon Inspired Optimization (MPIO) algorithm	UAVs	PIO DE PSO	• Evaluation function performance is better. • Convergence speed is enhanced. • Improved global searching process.	• Premature convergence issue. • Inaccuracy occurred due to large number of iterations.
7.	[16]	"PID Controller Design Based on Prey–Predator Pigeon-Inspired Optimization Algorithm"	Prey–Predator PIO algorithm (PPPIO)	Plant System	PIO PSO	• Improved PID tuning. • Avoid trapping optimal solution.	• Step response is similar to PIO. • Parameters values similar to PSO.

S. No	Ref	Title	Technique used	Applied on	Compared with	Key findings	Limitations
8.	[17]	"Bloch Quantum-behaved Pigeon-Inspired Optimization for Continuous Optimization Problems"	Bloch Quantum-behaved Pigeon Inspired Optimization (BQPIO)	Benchmark Function Problems	PSO QPSO PIO	• Achieved global minimum of the functions. • Optimal solution in less iteration.	• Valid for short reference only. • Non directional mutation operation caused slowed convergence
9.	[18]	"Gaussian pigeon-inspired optimization approach to orbital spacecraft formation reconfiguration"	Gaussian Pigeon Inspired Optimization (GPIO)	Orbital Spacecraft Formation	PSO PIO	• Simulation running time is smaller than PIO and PSO • Used Gaussian method for space craft formation reconfiguration	• The robustness value is just satisfied. • Achieved late fitness value than PSO and PIO
10.	[19]	"Pigeon-inspired optimization applied to constrained gliding trajectories"	Prey–Predator PIO algorithm (PPPIO)	Hypersonic Gliding Vehicles	PSO	• Generated the inhibited gliding route for hypersonic gliding vehicles. • Quick decision in flight's mission placement.	• Not feasible for uncertainties in complex environment • Results are same after 20 iterations.
11.	[20]	"Multi-objective pigeon-inspired optimization for brushless direct current motor parameter design"	Multi-objective PIO (MPIO)	BLDC motor	NSGA-II	• Improved pareto frontier obtained in parameter design. • Achieved more stability by introducing transition in operators.	• Inconsistence found in algorithm when used for BLDC
12.	[21]	"Robust Binocular Pose Estimation Based on Pigeon-Inspired Optimization"	PIO with Binocular pose estimation (PIO-BPnP)	Binocular Camera Systems	LHM MLHM RPnP BPnP	• Fine pose estimation received for binocular camera systems. • Minimized all types of errors. • Optimal results achieved in few iterations	• Too many simulations needed for average result • Inefficient because it needs all edge calculations.
13.	[22]	"Linear-quadratic regulator controller design for quadrotor based on Pigeon inspired optimization"	LQR based on PIO	Autonomous Aerial Refueling	PSO	• Best fitness value obtained • Estimated altitude achieved for quad rotor in least time.	• Energy cost increases due to constant R parameter.
14.	[23]	"A type of collective detection scheme with improved pigeon-inspired Optimization"	PIO with Expand and Contract concept (ECPIO) algorithm	Global Navigation Satellite Systems	PSO DE PIO	• Execution time is faster. • Complex multi-model functions outperformed performance of previous ones.	• Decision making process is missing • It's based on un real signals.

S. No	Ref	Title	technique used	applied on	compared with	key findings	limitations
15.	[24]	"Control parameter design for automatic carrier landing system via pigeon-inspired optimization"	Pigeon Inspired Optimization (PIO)	Automatic Carrier Landing System (ACLS)	BSO PSO ABC BA	• Expected pitch rate is achieved. • Best parameter tuning obtained for ACLS. • Fast convergence speed. • Minimum standard deviation.	• Initial parts are same when compared. • Only level 1 of control anticipation parameter is achieved.
16.	[25]	"Pendulum-like Oscillation Controller for UAV Based on Lévy-flight Pigeon-inspired Optimization and LQR"	Lévy Flight based PIO (LFPIO) algorithm for Linear-quadratic Regulator	UAVs	PIO PSO SAPSO	• More reliable in flight mode • Accuracy and convergence speed is much higher than PIO.	• Stability achieved after 20 iterations while other got before.
17.	[26]	"Pigeon inspired optimization approach to model prediction control for unmanned air Vehicles"	MPC controller with PIO algorithm	UAVs	PSO	• Best fitness value achieved in less than 10 iterations. • Reduced the parameter optimization burden from controller	• Step responses are same for PIO and PSO. • Still have capacity to improve in MPC controller.
18.	[27]	"Pigeon-inspired optimization and lateral inhibition for image matching of autonomous aerial refueling"	PIO with Lateral Inhibition (LI-PIO)	UAVs	PSO LI-PSO PIO	• Improved pre-processing images parameters i.e. contrast and edges. • Consume minimum time to execute	• PIO and LI-PIO have same stability 10th iteration • Results not included when the UAV tilted
19.	[28]	"A modified consensus algorithm for multi-UAV formation based on Pigeon inspired optimization with a slow diving strategy"	PIO with a Slow Diving Strategy	Multiple UAVs	PSO PIO	• Remove oscillations effectively and smooth the curve. • Achieved desired location with least fitness value.	• Sharp dive and quick climb may lead to crash. • Must remain at safe distance due to communication gap.
20.	[29]	"Active Disturbance Rejection Control for Small Unmanned Helicopters via Levy Flight-based Pigeon-inspired Optimization"	Lévy Flight pigeon-inspired optimization (LFPIO)	UAVs	PSO PIO	• Resolves altitude fluctuation problem in small unmanned helicopters.	• Step responses of angular and linear velocities approximately same.

S. No	Ref	Title	technique used	applied on	compared with	key findings	limitations
						• Optimized ADRC parameter to work best in complex environment.	• Small difference in lateral and longitudinal angular velocity step responses.
21.	[30]	"Aerodynamic Parameter Identification of Hypersonic Vehicle via Pigeon inspired Optimization"	Pigeon Inspired Optimization (PIO)	UAVs	PSO PIO	• Aerodynamic parameters identified with accuracy despite noise. • Minimum fluctuations in curve.	• PIO worked poor between 5 and 10 iterations. • Minimum difference in cost functions.
22.	[31]	"Biological object recognition approach using space variant resolution and pigeon-inspired optimization for UAV"	PIO with Space Variant Resolution mechanism (SVRPIO)	UAVs	GLPT-TM	• Computational complexity reduced due to optimized search approach. • With the help of rotational and scale challenges, object recognition is better.	• Efficiency is reduced when scaled twice. • Invalid run occurred when target moves in either side of coordinates simultaneously.
23.	[32]	"Flying Vehicle Longitudinal Controller Design via Prey–Predator Pigeon-Inspired Optimization"	Prey–Predator PIO algorithm (PPPIO)	Acceleration Control System	PIO PSO	• Faster response and no overshoot in designing a controller. • Improved normal acceleration performance	• Settling time is not suitable.
24.	[33]	"Fuzzy energy management strategy for parallel HEV based on pigeon-inspired optimization algorithm"	Quantum Chaotic Pigeon-Inspired Optimization (QCPIO) algorithm with Fuzzy approach.	Hybrid Electric Vehicle	Fuzzy PSO Fuzzy	• Reduced vehicle emission effectively • Improved fuel economy of vehicle. • Stable battery charging and discharging.	• Only applicable for low load area.
25.	[34]	"Lévy flight based pigeon-inspired optimization for control parameters optimization in automatic carrier landing system"	Lévy Flight based pigeon-inspired optimization	ACLS	PIO PSO DE	• Best landing track of aircraft in the presence of vertical wind disturbance. • Fewer fluctuations in angle of attack.	• DE lead in angle of attack with best fitness value for ACLS. • Took more time to remove the error

S. No	Ref	Title	technique used	applied on	compared with	key findings	limitations
26.	[35]	"Automatic Carrier Landing System multilayer parameter design based on Cauchy Mutation Pigeon-Inspired Optimization"	Cauchy Mutation Pigeon-Inspired Optimization (CMPIO)	ACLS	PSO DE PIO	• Improved bandwidth dynamic characteristics i.e. flight path response. • Smallest overshoot and least fitness function	• High rise and settling time.
27.	[36]	"Adaptive Operator Quantum-Behaved Pigeon-Inspired Optimization Algorithm with Application to UAV Path Planning"	Quantum-behaved pigeon-inspired optimization (QPIO)	UAVs	PSO PIO	• Smooth path planning in the presence of threat sources. • Convergence speed is faster	• Lower altitude value as compared to PSO and PIO. • Searching time is higher than PIO.
28.	[37]	"Social-class pigeon-inspired optimization and time stamp segmentation for multi UAV cooperative path planning"	Social Class PIO with Time Stamp Segmentation (SCPIO-TSS)	UAVs	PSO PIO	• Reduced coordination cost. • Explicit search optimization by modified landmark operator.	• Algorithm condition are not investigated • Slower convergence speed than PSO
29.	[38]	"Predator–Prey Pigeon-Inspired Optimization for UAV ALS Longitudinal Parameters Tuning"	predator–prey pigeon-inspired optimization (PPPIO) algorithm	UAVs	BBO, ES DE, PIO GA, StudGA	• Improved performance of flight path angle with nominal convergence speed. • Better integral value for ALS.	• Fluctuation in actual response of the proposed work. • Marginal stability exhibit by system.
30.	[39]	"A multi-objective pigeon-inspired optimization approach to UAV distributed flocking among obstacles"	multi-objective pigeon-inspired optimization (MPIO)	UAVs	MPIO NSGA-II	• Small population have fine Pareto frontier with few iterations. • Stable flight formation achieved	• Same curve for altitude and altitude rates. • Deadlock occurred in the convergence
31.	[40]	"Re-entry Trajectory Optimization using Pigeon Inspired Optimization Based Control Profiles"	Pigeon Inspired Optimization (PIO)	Spacecraft Launch Vehicles.	PSO	• Flew at upper bounds on load factor • Satisfactory entry trajectory as predicted	• Dynamic pressure of the system avoided. • Heart rate and initial flight path angle is slightly higher.
32.	[41]	"Mixed Game Pigeon-Inspired Optimization For Unmanned Aircraft System Swarm Formation"	Mixed Game Pigeon-Inspired Optimization (MGPIO)	Multiple UAVs	PIO PSO	• Stable formation with faster convergence. • System successfully avoided local minima.	• Chances of collision still present • Applied only on six UAV's.

S. No	Ref	Title	technique used	applied on	compared with	key findings	limitations
33.	[42]	"Coevolution Pigeon-Inspired Optimization with Cooperation-Competition Mechanism for Multi-UAV Cooperative Region Search"	coevolution pigeon-inspired optimization (CPIO) algorithm	multi-UAV cooperative search (MUCS)	PIO PSO GA	• Convergence speed is faster. • Best average number of target found among all	• Not suitable for large number of iterations
34.	[43]	"A pigeon-inspired optimization algorithm for many-objective optimization problems"	multi-objective pigeon inspired optimization (MPIO)	Multi Objective Optimization	NSGAIII GrEA HypE KnEA MOEA/D	• Stability and convergence speed is improved. • Best diversity achieved.	• Pareto fronts obtained from NSGA-III and MOEA/D are better than MaPIO.
35.	[44]	"Discrete pigeon-inspired optimization algorithm with Metropolis acceptance criterion for large-scale traveling salesman problem"	Discrete PIO (DPIO)	Float and Integer Distances	ESACO MAS SOM	• Improved performance of TSP. • Avoid premature convergence by modifying PIO basic operators.	• It cannot be implemented on large scale due to need of centralized access of data. • Need of fine-tuned parameters of DPIO.
36.	[45]	"Mobile Robot ADRC with an Automatic Parameter Tuning Mechanism via Modified Pigeon-inspired Optimization"	Evolutionary Game based PIO (EGPIO)	the deformable push rod	PIO PSO CPIO	• This method yields a minor steady-state error in overall angle output. • Settling time index is best among all.	• No convergence proof is given. • Kinematic limitations have been ignored in this method.
37.	[46]	"Dynamic Discrete Pigeon-inspired Optimization for Multi-UAV Cooperative Search-attack Mission Planning"	Dynamic Discrete PIO Algorithm (D^2PIO)	Multiple Unmanned Aerial Vehicles	PSO, BPSO, PIO,BPIO DPSO, MPSO	• Proposed method performed well as compared to others. • Task switching ability included in this research.	• Task completeness gradually decreasing due to frequent task switching. • Population size increment may lead to higher computational cost
38.	[47]	"Limit-Cycle-Based Mutant Multi-objective Pigeon-Inspired Optimization"	Limit-Cycle-based Mutant Multi-Objective Pigeon-Inspired Optimization (CMMOPIO)	Multi-Objective Optimization Algorithms	CMOPIO, MOPSO, NSGA-II, SPEA2, MOEA-D	• The faster convergence speed and avoid trapping into local optimum due to mutation mechanism. • Improved population diversity with wider search space.	• Real world problems • Performance of algorithm

S. No	Ref	Title	technique used	applied on	compared with	key findings	limitations
39.	[48]	"Multi-UAV obstacle avoidance control via multi-objective social learning pigeon-inspired optimization"	Multi-objective Social Learning Pigeon-Inspired Optimization (MSLPIO)	Unmanned Aerial Vehicle formation	MPIO NSGA-II	• Improved flocking control during flight. • Desired Yaw angle achieved • Great obstacle avoidance were seen	• Convergence speed seems poor • Some UAV's deviated from swarm when passed through obstacles.
40.	[49]	"A Binary Tree Based Coordination Scheme for Target Enclosing with Micro Aerial Vehicles"	Multi-UMAV target enclosing problem based on binary-tree communication topologies	Unmanned Micro Aerial Vehicle (UMAV)	UAV's	• Despite frequent variation in target direction proposed scheme is able to for stable formation. • Able to cover target at any nominal height.	• The method has some detection failure due to fixed communication among UAV's. • UMAV must have extraordinary flexibility.
41.	[50]	"A multi-strategy pigeon-inspired optimization approach to active disturbance rejection control parameters tuning for vertical take-off and landing fixed-wing UAV"	Multi-Strategy Pigeon-Inspired Optimization (MSPIO) algorithm	UAVs	PSO GA PIO CMPIO	• Solved the height fluctuation problem during forward flight of fixed wing UAV. • Hovering, dynamic inheritance to optimize better efficiency with improved search ability.	• Huge power required for transition. • No stability analysis have done.
42.	[51]	"A novel adaptive pigeon-inspired optimization algorithm based on evolutionary game theory"	adaptive pigeon-inspired optimization algorithm Evolutionary game theory (EGT)	pigeons	PIO CPIO CMPIO SCPIO	• Global optimization achieved with good convergence speed. • The system becomes stable after 50 iterations, which is good.	• Almost identical mean error curve for the Schwefel's function between CPIO and EGTPIO. • It does not support theoretical aspect
43.	[52]	"Autonomous trajectory tracking of a quadrotor UAV using ANFIS controller based on Gaussian pigeon-inspired optimization"	GPIO algorithm based on adaptive neuro-fuzzy inference system (ANFIS)	3-DOF quadrotor	PID	• Followed the reference trajectory at above 90% accuracy. • Convergence speed and stability also superior to classical method.	• Instability may occur in higher iterations

Table 1.
Comparative analysis of pigeon inspired optimization and its variants for motion planning.

Annealing Pigeon-inspired Optimization (SAPIO) and Edge Potential Function (EPF). Zhang and Duan [14] proposed a controller for formation reconfiguration problems of multiple unmanned aerial vehicles (UAVs). Hao et al. [15] linked PIO with energy consumption of UAV mission assignment. Sun and Duan [16] used PPPIO for Proportion-Integral-Derivative (PID) controller parameter adjustment. Li and Duan [17] proposed Bloch Quantum Behaved Pigeon-Inspired Optimization (BQPIO) to enhance the local search and position uncertainty.

In 2015, Shujian and Duan [18] presented another algorithm called improved pigeon-inspired optimization (PIO) algorithm of multiple orbital spacecraft formation problem. Jiang et al. [19] utilized PIO algorithm for the velocity-dependent bank angle profiles of the reentry vehicle. Hua et al. [20] used brushless DC motor parameters optimization via Multi-objective Pigeon Inspired Optimization (MPIO). Gan and Duan [21] presented a robust algorithm based on PIO for binocular pose estimation of multiple camera systems (MCS). Sun et al. [22] worked on PIO-based LQR controller for quad-rotor for autonomous aerial refueling (AAR). Zheng [23] proposed a new structure of CD for detection Global Navigation Satellite Systems (GNSS) signals and location by using improved pigeon-inspired optimization. Deng and Duan [24] presented a novel control parameter design method for the Automatic Carrier Landing System (ACLS) via PIO.

In 2016, Liu and Duan [25] developed a new Lévy -flight pigeon-inspired Optimization (LFPIO) algorithm for pendulum like oscillation controller in UAVs for optimality of LQR with accuracy, convergence speed and reliability. Dou and Duan [26] proposed PIO algorithm for parameter optimization in model prediction control (MPC) for unmanned air vehicles. Sun and Duan [27] showed a hybrid algorithm of lateral inhibition with pigeon inspired optimization (LI-PIO) autonomous aerial refueling (AAR) image matching problem.

In 2017, Zhang and Duan [28] proposed a new algorithm Slow Driving Strategy Pigeon Inspired Formula (SD-PIO) for Consensus. Zhang et al. [29] presented a novel algorithm LFPIO for active disturbance rejection control (ADRC) method applied on small unmanned helicopters. Xeu and Duan [30] opted PIO algorithm for aerodynamics parameters of hypersonic vehicles. Long and ning [31] proposed a novel global log-polar transformation (LPT) based template-matching algorithm (GLPT-TM) along with PIO for biological object recognition. Mohammad and Duan [32] developed Flying Vehicle Longitudinal Controller Design with the help of.

Prey–Predator Pigeon-Inspired Optimization (PPPIO), Zheng, et al. [33] proposed Quantum Chaotic Pigeon Inspired Optimization (QCPIO) algorithm for fuzzy control strategy of Hybrid Electric Vehicle (HEV). Dou and Duan [34] utilized a LFPIO for controlling the parameters of ACLS.

In 2018, Yang et al. [35] presents a novel algorithm Cauchy Mutation Pigeon Inspired Optimization (CMPIO) for the design problem of ACLS. Hu et al. [36] proposed Adaptive Operator Quantum-Behaved Pigeon-Inspired Optimization (AOQPIO) algorithm for UAV 3-D path planning problem. Zhang and Duan [37] proposed Social Class Pigeon Inspired Optimization (SCPIO) with Time Stamp Segmentation (TSS) for multi-UAV cooperative path planning. Duan et al. [38] used PPPIO optimization algorithm to improve the tracking control of the fixed-wing UAV. Qiu and Duan [39] applied MPIO for stable formation of UAV's in complex environment. Sushnigdha and Joshi [40] solved re-entry trajectory optimization problem of Spacecraft and launch vehicles by using PIO.

In 2019, Duan et al. [41] used Mixed Game Pigeon Inspired Optimization (MGPIO) algorithm for swarm formation of Unmanned Aircraft System (UAS). Luo et al. [42] proposed coevolution pigeon-inspired optimization (CPIO) algorithm for unmanned aerial vehicle (UAV) cooperative region search. Cui et al. [43] proposed a many-objective pigeon inspired optimization (MaPIOs) algorithm for multi-UAV cooperative region search. Zhong et al. [44] established discrete PIO (DPIO) algorithm for Traveling Salesman Problems (TSPs). Hai and Duan [45] proposed Evolutionary Game Theory based Pigeon Inspired Optimization (EGPIO) for autonomous mobile robot to boost ADRC method for the attitude deformation system.

In 2020, Duan et al. [46] proposed a Dynamic Discrete Pigeon Inspired Optimization (DDPIO) algorithm to solve a mission planning problem of search and attack of multiple UAVs. Duan et al. [47] presented Limit-Cycle-based Mutant Multi-Objective Pigeon-Inspired Optimization (CMMOPIO) to balance the global exploration and local exploitation. Ruan and Duan [48] proposed an improved PIO namely Multi-objective Social Learning Pigeon-Inspired Optimization (MSLPIO) for obstacle avoidance problem of Multi-UAV. Duan and Zhang [49] proposed coordination scheme for target enclosing based on binary tree for MUAV's.

In 2021, He and Duan [50] used a Multi-Strategy Pigeon-Inspired Optimization (MSPIO) algorithm to employ ADRC fluctuation problem HAI, et al. [51] utilized EGPIO algorithm to increase accuracy among pigeons. Selma et al. [52] mixed ANFIS controller with Gaussian pigeon-inspired optimization for autonomous trajectory tracking of a quad rotor UAV.

Above discussion is based on the improvements and modifications of basic PIO algorithms in each corresponding year. It can be seen that each year PIO, its modification and hybrid model become top trend in optimization related issues especially for the motion planning of various agents. For hybrid models, combination of other bio-inspired algorithm like ACO, GA etc. with PIO still lacking in this area.

5. Conclusion

Today, optimization algorithms are being widely used for the motion planning of complex optimization problems i.e. clusters, swarms and multi-objective by research scholars. Mostly, bio-inspired algorithms along with its variants have been proposed to increase the convergence speed and overall stability of the system. A novel bio-inspired optimization algorithm namely Pigeon Inspired Algorithms and its hybrid models are outperforming other related algorithm in terms of optimal motion planning techniques. This article manipulates recent trends of Pigeon Inspired Optimization algorithm and its modification for motion planning problems of agents. The dominance of PIO along with its hybrid model, an estimation mechanism must be developed in order to point of the importance over other bio inspired optimization algorithms. This study will help researcher to choose proper PIO variant for unexplored problem identification in complex environment where other known algorithm becomes failure. For future work, application based review or survey might be suitable for readers with hybrid model approach. Also work can be split into many parts based on path planning, formation control and self-organization of distributed systems.

Author details

Muhammad Shafiq[1], Zain Anwar Ali[1*] and Eman H. Alkhammash[2]

1 Electronic Engineering Department, Sir Syed University of Engineering and Technology, Karachi, Pakistan

2 Department of Computer Science, College of Computers and Information Technology, Taif University, Taif, Saudi Arabia

*Address all correspondence to: zainanwar86@hotmail.com

References

[1] Wiltschko, Wolfgang, and Roswitha Wiltschko. "Homing pigeons as a model for avian navigation?." Journal of Avian Biology 48, no. 1 (2017): 66-74.

[2] Keeton, William T. "The mystery of pigeon homing." Scientific American 231, no. 6 (1974): 96-107.

[3] Walcott, Charles. "Magnetic orientation in homing pigeons." IEEE Transactions on Magnetics 16, no. 5 (1980): 1008-1013.

[4] Guilford, Tim, Stephen Roberts, Dora Biro, and Iead Rezek. "Positional entropy during pigeon homing II: navigational interpretation of Bayesian latent state models." Journal of theoretical biology 227, no. 1 (2004): 25-38.

[5] Nießner, Christine, Susanne Denzau, Leo Peichl, Wolfgang Wiltschko, and Roswitha Wiltschko. "Magnetoreception in birds: I. Immunohistochemical studies concerning the cryptochrome cycle." Journal of Experimental Biology 217, no. 23 (2014): 4221-4224.

[6] Biro, Dora, Tim Guilford, Giacomo Dell'Omo, and Hans-Peter Lipp. "How the viewing of familiar landscapes prior to release allows pigeons to home faster: evidence from GPS tracking." Journal of Experimental Biology 205, no. 24 (2002): 3833-3844.

[7] Hagstrum, Jonathan T. "Atmospheric propagation modeling indicates homing pigeons use loft-specific infrasonic 'map'cues." Journal of Experimental Biology 216, no. 4 (2013): 687-699.

[8] Katzung Hokanson, Brandon R. "Saving grace on feathered wings: homing pigeons in the first world war." The Gettysburg Historical Journal 17, no. 1 (2018): 7.

[9] Duan, Haibin, and Huaxin Qiu. "Advancements in pigeon-inspired optimization and its variants." Sci. China Inf. Sci. 62, no. 7 (2019): 70201-1.

[10] Duan, Haibin, and Peixin Qiao. "Pigeon-inspired optimization: a new swarm intelligence optimizer for air robot path planning." International journal of intelligent computing and cybernetics (2014).

[11] Zhang, Bo, and Haibin Duan. "Predator-prey pigeon-inspired optimization for UAV three-dimensional path planning." In International Conference in Swarm Intelligence, pp. 96-105. Springer, Cham, 2014.

[12] Zhang, Bo, and Haibin Duan. "Three-dimensional path planning for uninhabited combat aerial vehicle based on predator-prey pigeon-inspired optimization in dynamic environment." IEEE/ACM transactions on computational biology and bioinformatics 14, no. 1 (2015): 97-107.

[13] Li, Cong, and Haibin Duan. "Target detection approach for UAVs via improved pigeon-inspired optimization and edge potential function." Aerospace Science and Technology 39 (2014): 352-360.

[14] Zhang, Xiaomin, Haibin Duan, and Chen Yang. "Pigeon-inspired optimization approach to multiple UAVs formation reconfiguration controller design." In Proceedings of 2014 IEEE Chinese Guidance, Navigation and Control Conference, pp. 2707-2712. IEEE, 2014.

[15] Hao, Ran, Delin Luo, and Haibin Duan. "Multiple UAVs mission assignment based on modified pigeon-inspired optimization algorithm." In Proceedings of 2014 IEEE Chinese Guidance, Navigation and Control Conference, pp. 2692-2697. IEEE, 2014.

[16] Sun, Hang, and Haibin Duan. "PID controller design based on prey-predator pigeon-inspired optimization algorithm." In 2014 IEEE international conference on mechatronics and automation, pp. 1416-1421. IEEE, 2014.

[17] Li, Honghao, and Haibin Duan. "Bloch quantum-behaved Pigeon-inspired optimization for continuous optimization problems." In Proceedings of 2014 IEEE Chinese Guidance, Navigation and Control Conference, pp. 2634-2638. IEEE, 2014.

[18] Zhang, Shujian, and Haibin Duan. "Gaussian pigeon-inspired optimization approach to orbital spacecraft formation reconfiguration." Chinese Journal of Aeronautics 28, no. 1 (2015): 200-205.

[19] Zhao, Jiang, and Rui Zhou. "Pigeon-inspired optimization applied to constrained gliding trajectories." Nonlinear dynamics 82, no. 4 (2015): 1781-1795.

[20] Qiu, HuaXin, and HaiBin Duan. "Multi-objective pigeon-inspired optimization for brushless direct current motor parameter design." Science China Technological Sciences 58, no. 11 (2015): 1915-1923.

[21] Gan, Lu, and Haibin Duan. "Robust binocular pose estimation based on pigeon-inspired optimization." In 2015 IEEE 10th Conference on Industrial Electronics and Applications (ICIEA), pp. 1043-1048. IEEE, 2015.

[22] Sun, Yongbin, Ning Xian, and Haibin Duan. "Linear-quadratic regulator controller design for quadrotor based on pigeon-inspired optimization." Aircraft Engineering and Aerospace Technology (2016).

[23] Zhengxuan, J. I. A. "A type of collective detection scheme with improved pigeon-inspired optimization." International Journal of Intelligent Computing and Cybernetics (2016).

[24] Deng, Yimin, and Haibin Duan. "Control parameter design for automatic carrier landing system via pigeon-inspired optimization." Nonlinear Dynamics 85, no. 1 (2016): 97-106.

[25] Liu, Zhuqing, Haibin Duan, Yijun Yang, and Xiaoguang Hu. "Pendulum-like oscillation controller for UAV based on Lévy-flight pigeon-inspired optimization and LQR." In 2016 IEEE Symposium Series on Computational Intelligence (SSCI), pp. 1-6. IEEE, 2016.

[26] Dou, Rui, and Haibin Duan. "Pigeon inspired optimization approach to model prediction control for unmanned air vehicles." Aircraft Engineering and Aerospace Technology: An International Journal (2016).

[27] Sun, Yongbin, and Haibin Duan. "Pigeon-inspired optimization and lateral inhibition for image matching of autonomous aerial refueling." Proceedings of the Institution of Mechanical Engineers, Part G: Journal of Aerospace Engineering 232, no. 8 (2018): 1571-1583.

[28] Zhang, Tianjie, and Haibin Duan. "A modified consensus algorithm for multi-UAV formations based on pigeon-inspired optimization with a slow diving strategy." J Intell Syst (in China) 12, no. 4 (2017): 570-581.

[29] Zhang, Daifeng, Haibin Duan, and Yijun Yang. "Active disturbance rejection control for small unmanned helicopters via levy flight-based pigeon-inspired optimization." Aircraft Engineering and Aerospace Technology (2017).

[30] Xue, Qiang, and Duan Haibin. "Aerodynamic parameter identification of hypersonic vehicle via pigeon-inspired optimization." Aircraft Engineering and Aerospace Technology (2017).

[31] Xin, Long, and Ning Xian. "Biological object recognition approach

using space variant resolution and pigeon-inspired optimization for UAV." Science China Technological Sciences 60, no. 10 (2017): 1577-1584.

[32] Mohamed, Mostafa S., Haibin Duan, and Li Fu. "Flying vehicle longitudinal controller design via prey-predator pigeon-inspired optimization." In 2017 IEEE Symposium Series on Computational Intelligence (SSCI), pp. 1-6. IEEE, 2017.

[33] Pei, JiaZheng, YiXin Su, and DanHong Zhang. "Fuzzy energy management strategy for parallel HEV based on pigeon-inspired optimization algorithm." Science China Technological Sciences 60, no. 3 (2017): 425-433.

[34] Dou, Rui, and Haibin Duan. "Lévy flight based pigeon-inspired optimization for control parameters optimization in automatic carrier landing system." Aerospace Science and Technology 61 (2017): 11-20.

[35] Yang, Zhiyuan, Haibin Duan, Yanming Fan, and Yimin Deng. "Automatic carrier landing system multilayer parameter design based on cauchy mutation pigeon-inspired optimization." Aerospace Science and Technology 79 (2018): 518-530.

[36] Hu, Chunhe, Yu Xia, and Junguo Zhang. "Adaptive operator quantum-behaved pigeon-inspired optimization algorithm with application to UAV path planning." Algorithms 12, no. 1 (2019): 3.

[37] Zhang, Daifeng, and Haibin Duan. "Social-class pigeon-inspired optimization and time stamp segmentation for multi-UAV cooperative path planning." Neurocomputing 313 (2018): 229-246.

[38] Duan, Haibin, Mengzhen Huo, Zhiyuan Yang, Yuhui Shi, and Qinan Luo. "Predator-prey pigeon-inspired optimization for UAV ALS longitudinal parameters tuning." IEEE Transactions on Aerospace and Electronic Systems 55, no. 5 (2018): 2347-2358.

[39] Qiu, Huaxin, and Haibin Duan. "A multi-objective pigeon-inspired optimization approach to UAV distributed flocking among obstacles." Information Sciences 509 (2020): 515-529.

[40] Sushnigdha, Gangireddy, and Ashok Joshi. "Re-entry trajectory optimization using pigeon inspired optimization based control profiles." Advances in Space Research 62, no. 11 (2018): 3170-3186.

[41] Duan, Haibin, Bingda Tong, Yin Wang, and Chen Wei. "Mixed game pigeon-inspired optimization for unmanned aircraft system swarm formation." In International Conference on Swarm Intelligence, pp. 429-438. Springer, Cham, 2019.

[42] Luo, Delin, Jiang Shao, Yang Xu, Yancheng You, and Haibin Duan. "Coevolution pigeon-inspired optimization with cooperation-competition mechanism for multi-UAV cooperative region search." Applied Sciences 9, no. 5 (2019): 827.

[43] Cui, Zhihua, Jiangjiang Zhang, Yechuang Wang, Yang Cao, Xingjuan Cai, Wensheng Zhang, and Jinjun Chen. "A pigeon-inspired optimization algorithm for many-objective optimization problems." Sci. China Inf. Sci. 62, no. 7 (2019): 70212-1.

[44] Zhong, Yiwen, Lijin Wang, Min Lin, and Hui Zhang. "Discrete pigeon-inspired optimization algorithm with Metropolis acceptance criterion for large-scale traveling salesman problem." Swarm and Evolutionary Computation 48 (2019): 134-144.

[45] Hai, Xingshuo, Zili Wang, Qiang Feng, Yi Ren, Binghui Xu, Jingjing Cui, and Haibin Duan. "Mobile robot ADRC with an automatic parameter tuning

mechanism via modified pigeon-inspired optimization." IEEE/ASME Transactions on Mechatronics 24, no. 6 (2019): 2616-2626.

[46] Duan, Haibin, Jianxia Zhao, Yimin Deng, Yuhui Shi, and Xilun Ding. "Dynamic Discrete Pigeon-Inspired Optimization for Multi-UAV Cooperative Search-Attack Mission Planning." IEEE Transactions on Aerospace and Electronic Systems 57, no. 1 (2020): 706-720.

[47] Duan, Haibin, Mengzhen Huo, and Yuhui Shi. "Limit-cycle-based mutant multiobjective pigeon-inspired optimization." IEEE Transactions on Evolutionary Computation 24, no. 5 (2020): 948-959.

[48] Ruan, Wan-ying, and Hai-bin Duan. "Multi-UAV obstacle avoidance control via multi-objective social learning pigeon-inspired optimization." Frontiers of Information Technology & Electronic Engineering 21 (2020): 740-748.

[49] Duan, Haibin, and Daifeng Zhang. "A Binary Tree Based Coordination Scheme for Target Enclosing with Micro Aerial Vehicles." IEEE/ASME Transactions on Mechatronics 26, no. 1 (2020): 458-468.

[50] He, Hangxuan, and Haibin Duan. "A multi-strategy pigeon-inspired optimization approach to active disturbance rejection control parameters tuning for vertical take-off and landing fixed-wing UAV." Chinese Journal of Aeronautics (2021).

[51] Hai, Xingshuo, Zili Wang, Qiang Feng, Yi Ren, Bo Sun, and Dezhen Yang. "A novel adaptive pigeon-inspired optimization algorithm based on evolutionary game theory." Science China Information Sciences 64 (2021): 1-2.

[52] Selma, Boumediene, Samira Chouraqui, Belkacem Selma, Hassane Abouaïssa, and Toufik Bakir. "Autonomous trajectory tracking of a quadrotor UAV using ANFIS controller based on Gaussian pigeon-inspired optimization." CEAS Aeronautical Journal 12, no. 1 (2021).

Section 3

Wheeled Robots Planning and Control

Chapter 5

Autonomous Vehicle Path Planning Using MPC and APF

Zahra Elmi and Soheila Elmi

Abstract

Autonomous vehicles have been at the forefront of academic and industrial research in recent decades. This study's aim is to reduce traffic congestion, improve safety, and accidents. Path planning algorithms are one of the main elements in autonomous vehicles that make critical decisions. Motion planning methods are required when transporting passengers from one point to another. These methods have incorporated several methods such as generating the best trajectory while considering the constraints of vehicle dynamics and obstacles, searching a path to follow, and avoiding obstacles that guarantee comfort, safety, and efficiency. We suggested an effective path planning algorithm based on Model Predictive Controller that determines the maneuvers mode such as lane-keeping and lane-changing automatically. We utilized two different artificial potential field functions for the road boundary, obstacles, and lane center to ensure safety. On the four scenarios, we examined the proposed path planning controller. The obtained results show that when a path planning controller is used, the vehicle avoids colliding with obstacles and follows the rules of the road by adjusting the vehicle's dynamics. An autonomous vehicle's safety is ensured by the path planning controller.

Keywords: motion planning, model predictive controller, potential field, autonomous vehicle, obstacle avoidance

1. Introduction

Today, autonomous robots can be utilized in a number of roles in our daily life. The autonomous robots without human intervention are able to move in the environment and perform their tasks safely and have a wide variety of applications. The main goals are to help humans with difficult, repetitive, and tedious tasks. Additionally, substituting the robots for humans in these tasks is an important dream of humans to reduce human-based errors. Therefore, many developments have done in term of software, hardware, computing, and control. One of the important techniques in robotic science is related to path planning that the goal is to plan a path with the movement of the robot from a start position to target while avoiding collisions with static and dynamic obstacles in the environment. Path planning is a challenging decision-making and control problem. This problem performs in two ways: first, global path planning that the knowledge of the environment is fully available for robot and robot is able to reach to target position safely. Second, local path planning is performed using only the sensed data by the robot, namely, the knowledge of the environment is unidentified or partially unidentified.

Many path planning approaches are presented, which can be divided into two categories: conventional and heuristic approaches. Common methods such as Roadmap [1], Potential Field [2], Cell Decomposition [3], and Mathematical Programming are examples of conventional approaches. These methods are used as hybridization in many applications. Heuristic approaches are presented to overcome the limitations of conventional methods. Probabilistic Roadmap [4], Simulated Annealing [5], Ant Colony Optimization [6], Particle Swarm Optimization [7], and Grasshopper optimization [8] are a few examples of heuristic methods. However, these algorithms have problems in static and dynamic environments. One of the simplest heuristic algorithms is the Dijkstra algorithm that is based on graph search and is able to find a minimum path between two different nodes on a graph by discretization of the environment. The other algorithm is A* which is similar to the Dijkstra algorithm but uses two cost functions to move from start position to target. These algorithms are applied only for environments with static obstacles. These algorithms guaranteed efficiency and optimality of obtaining path but the planned path depends on the resolution of the graph highly. Moreover, considering the dynamic constraints of robots is difficult during the planning process. In the enhanced version of A*, the authors [9] used the proposed method in a changeable environment, and the result is shown that the planned and tracked path is smoother than traditional methods. However, this method ignores the dynamic constraints of obstacles.

The planners of curve interpolation such as Clothoid, Polynomial, Spline, and Bezier curves are widely used for online path planning. These planners are similar to methods based on graph search and have low computational cost because the behavior of the curve is defined by a few control points or parameters. However, the optimality of obtained path is not guaranteed, and the dynamic constraints of a robot are not considered during the planning process and are additionally needed a smoothing process for the obtained path. In [10], a new method is presented by the Clothoid curve to reduce the length and curvature change of path. In this approach, two points are considered on the plane and the proposed algorithm generates a closed-form solution to connect two Clothoid sets for the position of a waypoint. This approach reduces sudden changes of curvature and sideslip by robot and improves the performance of the movement. To generate trajectory in [11], the authors used the polynomial parameterization that represents kinematic constraints and moving obstacles. Besides, the velocity of the robot is planned using this parameterization. To find the optimal solution, a guideline obtained by the Bezier curve is introduced. The result of the simulation has shown that the proposed method performs better than the traditional one. In [12], the authors used a combination of RRT* and Spline techniques to generate a smooth path. The proposed bidirectional Spline-RRT* algorithm is based on the cubic curve and satisfies direction constraints for both start and target positions. This algorithm is not similar to other path planning algorithms and the obtained result for the robot is sub-optimal yet feasible.

Some of the heuristic algorithms such as simulated annealing [13, 14] are used for path planning in environments with static and dynamic obstacles. This algorithm improves the obtained path for the robot and processing epoch. The obtained path is a near-optimal solution and is possible for online implementation, but it ignores the dimension of the robot and avoids only obstacles with circular shapes. Probabilistic Road Maps and Rapidly Random Tress are considered as the methods based on sampling. These methods are used for both holonomic and non-holonomic systems. In [15], a hybrid method for navigation of robots in dynamic and unknown environments is proposed. The proposed method is a combination of the proposed reactive planner and a global planner that Dynamic Rapidly exploring Random Tree

(DRRT) algorithm is used as a global planner. This method improves the speed of planning by reusing parts of the old tree when re-growing it. A probabilistic local planner is used to avoid obstacles. The result illustrates that the proposed method has a 77% reduction in the configured environments. The RRT* [16] guarantees the optimality of obtained path and can achieve convergence to global optimal solution by increasing the number of samples. These methods and their variants are widely used for autonomous robot research. But it is impossible to use in practical applications since they have high computational complexity.

In methods based on path optimization, the main idea is to formulate path planning as an optimization problem that the desired performance and constraints of the robot are considered. This approach is able to find a proper path between start and target positions. In [17], a novel method is presented to predict and avoid the collision of static and dynamic obstacles in an unknown environment. To predict the velocity of obstacles, they have used a decision-making process by using the information of the sensory system of the robot. Therefore, the robot is able to find the proper path, reach the target safely and without any collision. The result illustrates the efficient algorithm for complex and dynamic environments. In [18], an uncontrollable divergence metric is presented. A mechanism to switch between multiple predictive controllers is developed by using this metric to reduce the return time of the controller and maintain predictive accuracy. In [19], a nonlinear MPC for an autonomous underwater vehicle (AUV) is offered. The path planning problem is solved with a receding horizon optimization framework with a Spline template. A combination of the obtained result from path planning and MPC is used for tracking control. To determine the maneuvers mode for autonomous vehicles in dynamic environments, a path planning method with MPC is proposed [20]. To decide maneuvers of lane change and lane-keeping, the convex relaxation method is used. For ensuring the safety of vehicles, a collision-avoidance method is developed. Also, for having a comfortable and natural maneuver, the lane-associated potential field is presented.

The contribution of this paper is to develop a nonlinear MPC approach to solve the path planning problem of an autonomous vehicle. The rest of the paper is organized as follows. The overall framework of an autonomous vehicle and the artificial potential field functions for the road and obstacles, and the model predictive controller for path planning are introduced in Section 2. Section 3 evaluates and discusses the results of path planning for four scenarios. Finally, conclusion is provided in Section 4.

2. Problem description

This section describes the vehicle framework that was used for simulation and control design. **Figure 1** depicts the autonomous vehicle highway scenario used in this paper. The main goal of this paper is to transport a vehicle from a given origin to a given destination at a controlled speed while adhering to common traffic rules

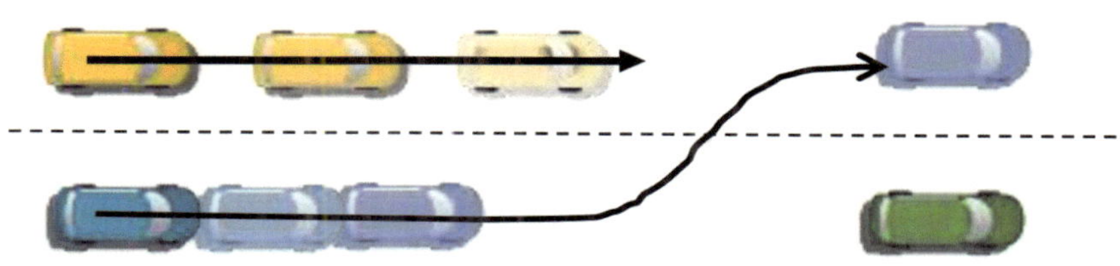

Figure 1.
The structure of driving environment and the state of obstacle.

and conventions, as well as to avoid colliding with obstacles and to provide a pleasant driving experience. As mentioned earlier, mathematical optimization methods have recently been found to be interesting. These methods provide a symmetric and precise method for considering vehicle dynamics and safety constraints, and they generate the optimal control inputs as a result.

If the environment is fully pre-identified, a mathematical optimization method is used in open-loop form; if the environment is unidentified, a feedback controller is utilized to recognize it [21–22]. In most research, MPC is one of the mathematical optimization approaches applied for online path planning.

Since the driving environment is usually dynamic and stochastic, and cannot be fully predicted a priori, the Model Predictive Control (MPC) approach for path planning has become popular in recent years. MPC uses a recursive method to synthesize a sequence of control optimal inputs in a finite time. The state of the robot or vehicle is updated according to this sequence [23–26]. Two different levels of the controller are considered to solve the path planning problem in a dynamic environment, as shown in **Figure 2**. The first level is for path planning, which generates a reference path based on environmental and destination data. The other controller is for the tracking path, which tracks the reference path directly using control inputs. The path planner employs a kinematic model of the vehicle, while the path tracker employs a dynamic model. The main goal of this paper is to develop a nonlinear MPC approach to solve the path planning problem for autonomous vehicles. The vehicle can plan its path over a finite horizon by using MPC.

2.1 The framework of the vehicle

a bicycle model is used to model the vehicle dynamics. In this model, the vehicle's two front wheels are combined into a single wheel in the front axle's center, while the vehicle's two rear wheels are in the rear axle's center. The kinematic framework is used to model the ego vehicle as well as any obstacles or other vehicles in the area. In the meantime, **Figure 3** depicts the vehicle model.

The motion equations of the bicycle model are as follows:

$$\ddot{x} = \dot{y}\theta + a_x$$

$$m\ddot{y} = 2\left[C_{\alpha f}\left(\delta_f - \frac{\dot{y} + l_f\theta}{\dot{x}}\right) + C_{\alpha r}\frac{l_r\theta - \dot{y}}{\dot{x}})\right] - m\dot{x}\theta$$

$$\dot{\varphi} = \omega$$

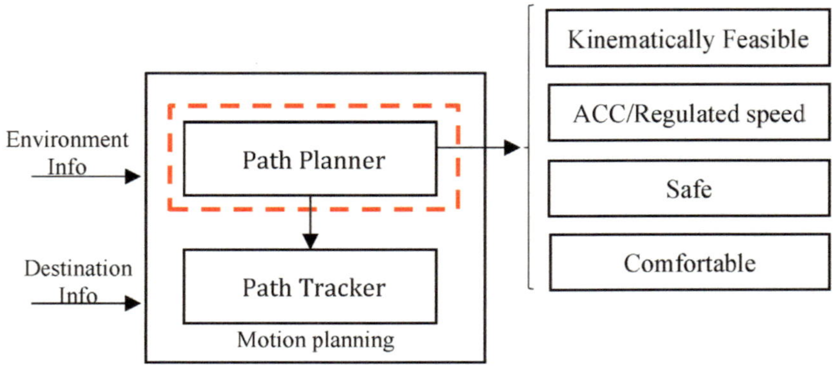

Figure 2.
The motion planning for the autonomous vehicle.

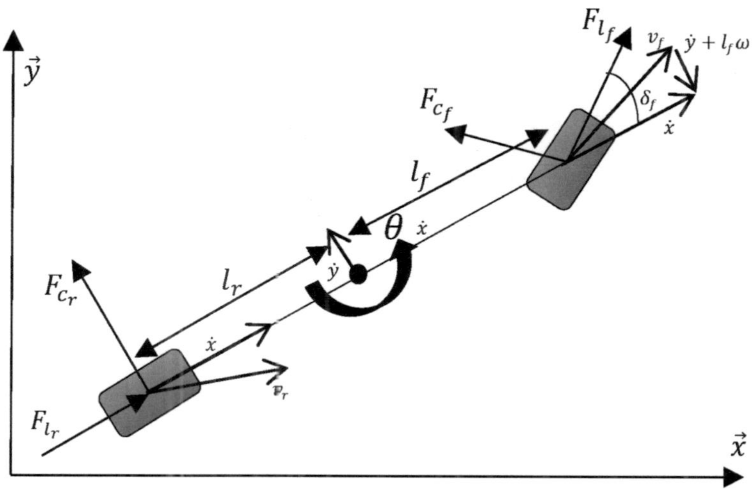

Figure 3.
The model of bicycle vehicle.

$$
\begin{gathered}
I_z\dot{\theta} = 2\left[l_f C_{\alpha f}\left(\delta_f - \frac{\dot{y}+l_f\theta}{\dot{x}}\right) - l_{fr} C_{\alpha r}\frac{l_r\theta - \dot{y}}{\dot{x}})\right] \\
\dot{X} = \dot{x}cos\varphi - \dot{y}sin\varphi \\
\dot{Y} = \dot{x}sin\varphi - \dot{y}cos\varphi
\end{gathered} \tag{1}
$$

where the vehicle's longitudinal and lateral velocities are$\dot{x}$ and $\dot{y}$. The yaw rate and yaw angle of the vehicle are φ and θ. X and Y are the longitudinal and lateral positions. The front steering angle and the vehicle longitudinal acceleration are represented by δ_f and a_x. The distances between the front and rear axles and the vehicle center of gravity are denoted by l_f and l_r. The cornering stiffness of the front and rear tires are indicated by $C_{\alpha f}$and $C_{\alpha r}$. The mass and inertia moment of the vehicle are m and I_z.

The model is linearized at an operational point and updated to the MPC internal prediction model at each control step in order to apply a non-linear model in linear MPC. Furthermore, the zero-order-hold technique is used to discretize the linearized model that can be obtained as follows:

$$
\begin{gathered}
\xi(t+1) = A(t)x(t) + B(t)u(t) \\
\xi = (\dot{x}, \dot{y}, \varphi, \theta, X, Y)u = \left(a_x, \delta_f\right)
\end{gathered} \tag{2}
$$

where ξ and u are vectors of state and input, respectively. State and input matrices are represented by A and B.

2.2 Artificial potential field function (APF)

The attractive and repulsive functions in the potential field (PF) technique allow the vehicle to proceed towards the objective, while the repulsive function prevents the vehicle from colliding with obstacle vehicles. The target potential field attracts the vehicle since it has a minimum value in the target location however, the obstacle potential field function repulses the vehicle from the obstacle because it has a maximum value in the obstacle locations [27]. The major goal of this paper is the navigation of a vehicle to a target point without colliding, which is accomplished by tracking an objective function. As a result, we just regard repulsive function as potential field. For this reason, the obstacle PF (U_O)and the road boundaries (U_R) are used to build the potential field function. At each prediction time, the total sum of potential field functions is obtained by reflecting the predicted surrounding environment. Obstacle vehicles are predicted as a model with constant velocity, and also the information of these vehicles is taken into account in real-time. The sum of the PFs is the potential field:

$$U_{tot} = \lambda_r U_R + \lambda_o U_O \quad (3)$$

where λ_r and λ_o are weights of PF for road and obstacle, respectively. To model road regulation and obstacles, Other functions can be also offered.

2.2.1 PF of lane marker

The lane marker PF is used to keep the vehicle from leaving the main route and driving too close to the boundaries of the road, which leads to raising the risk of a crash. As a result, the lane marker on the road borders should have a maximum value. Furthermore, the slope of achieving this peak point is maximum, so enables a restoring force of maximum value. Meantime, this peak point is operated at the position of the driving lane to prevent changing lanes. Hence, the vehicle tries to keep its current lane to avoid incurring further costs. For this reason, when the vehicle is not facing traffic or barriers, in the center of the lane, PF is zero and locally symmetric that is desired location. The vehicle can overcome this barrier when changing lanes is required. As a result, we utilize a 1D Gaussian function that approaches the left or right road border to get a larger potential value. The following is the PF for the lane marker (U_R):

$$U_R = A_r \exp\left(-\frac{(Y_h - Y_r)^2}{2\sigma_{rb}^2}\right) + A_r \exp\left(-\frac{(Y_h - Y_l)^2}{2\sigma_{rb}^2}\right) \quad (4)$$

where A_r is the maximum value of the potential field for the road boundary. Y_h denotes the lateral position of the ego vehicle in the local road frame, whereas Y_r, Y_l denote the lateral locations of the ego vehicle to the right and left of the center of the straight road, respectively.σ_{rb} is the variable for the road boundary's potential field. In this paper, we assume that the autonomous vehicle is driving on highway then the geometric shape of the road boundary is considered as a first order polynomial function. **Figure 4(a)** shows the 3D plot of the potential filed of road boundary on the straight road with $Y_r = -3.8$, $Y_l = 3.8$, $A_r = 40$ and σ_{rb}=1.

2.2.2 Obstacle potential field

The obstacle PF (U_O) framework is more complicated and essential than the road PF structure. According to the obstacle PF, the lane change movement is performed if the obstacle vehicle approaches the ego vehicle. This is based on highway driving's structure and protocol. In addition, the vehicle may shift to the left side to pass slower preceding vehicles. For accomplishing this, obstacle PF is modeled as a function of the measured position of the obstacle vehicle, relative and absolute velocity of the vehicle, road curvature, and obstacle vehicles. The available sensor readings from the obstacle are used to determine the location of the obstacle PF. The longitudinal and lateral distances between the ego vehicle and obstacle,

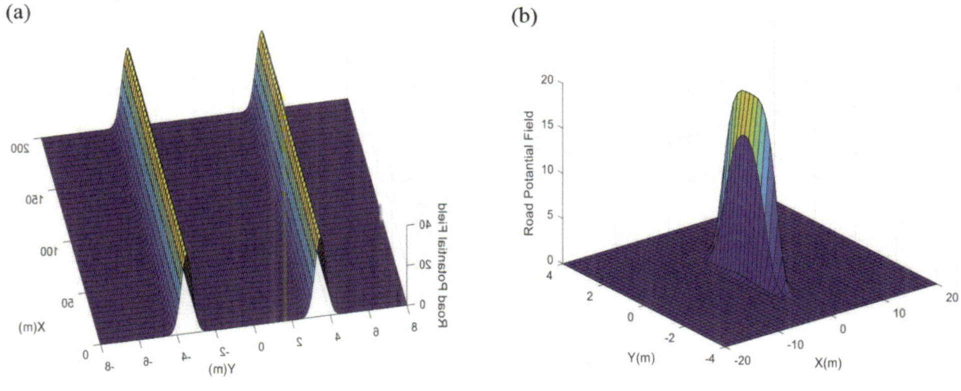

Figure 4.
The potential field function (a) road boundary (b) obstacle or surrounding vehicle.

given by x_O and y_O, are the acquired information and do not contain the obstacle vehicle's heading angle.

Since it gives a better representation of the layout of an obstacle, the form of an obstacle vehicle is rectangular. To prevent slope discontinuities in PF, continuous functions, such as the hyperbolic function, must be used to describe the obstacle value. This function produces the required potential field by determining the distance between the ego and obstacle vehicles. The function's change rate is rigorously raised when the distance between the ego vehicle and the obstacle is too tiny, and its value approaches infinity, preventing the ego vehicle from colliding with the obstacle. The obstacle's repulsive potential function is as follows:

$$U_o = A_{obs} \exp\left(-\left(\frac{(x - x_{obs})^2}{2\sigma_x^2} + \frac{(y - y_{obs})^2}{2\sigma_y^2}\right)^c\right) \tag{5}$$

where A_{obs} shows the maximum potential field value of the obstacle. (x, y) is the current position of the vehicle and (x_{obs}, y_{obs}) represents the nearest point of ego vehicle from the obstacle. σ_x and σ_y are the convergence coefficient of the obstacle potential field that determines the spread of the horizontal influence of the potential field. In Eq. (5), c is a coefficient for adjusting the shape of the obstacle potential field's peak. The approaching velocity is equal to the difference of velocity between them, If the ego and obstacle are closing together in each direction otherwise, it is set to zero. (x_{obs}, y_{obs}) = (0,0) is the location of the possible obstacle field, also $\sigma_x = \sigma_y = 1$ and A_{obs} = 20 are considered and obstacle potential field is shown in **Figure 4(b)**.

2.2.3 MPC framework

MPC is a hybrid method that combines optimum and adaptive control systems [28]. The approach employs a controller-based model that is used in the optimization stage of the model's anticipated states in order to provide the best control input. As a result, the MPC is comparable to an adaptive controller in that it can respond to changing conditions. It manages input and output restrictions at each control interval to solve the optimization problem. MPC is a viable alternative for path planning and tracking based on these characteristics' potential fields. Based on a dynamic model of the vehicle, road regulations and potential field functions, a model predictive controller is suggested. An optimization problem with conflicting needs can be defined using these objectives. The model predictive controller predicts the response of the ego vehicle based on a horizon known as the prediction horizon (N) and optimizes the vehicle's response, obstacle avoidance, road regulation, and command following based on this value. The intended lane and speed are predefined. Hence, the desired lateral position (the center of the desired lane) and longitudinal velocity are the system outputs that should be tracked:

$$y = [Y\ v_x]^T$$
$$y_{des} = [Y_{des} v_{xdes}]^T$$
$$Y_{des} = \left(l_{des} - \frac{1}{2}\right)L_w + \Delta Y_R \tag{6}$$

where y is the output matrix tracking, y_{des} is the intended lateral position, v_{xdes} is the desired speed, l_{des} is the index number of the desired lane from the right, L_w is the lane width, and ΔY_R is the lateral offset of the road relative to a straight road. The path planning nonlinear optimization problem can be expressed in the form:

$$\min_{u,s} \sum_{i=1}^{N} \|y(t+i|t) - y_{des}(t+i|t)\|_Q^2 + \|u(t+i-1|t) - u(t+i-|t)\|_R^2 + \|\mathrm{u}(t+i-1|t)\|_S^2 + U_R(t+i|t) + U_O(t+i|t) + \|s_i\|_P^2 \tag{7}$$

s.t.

$$x(t+i|t) = x(t+i-1|t) + u(t+i-1|t) \tag{8}$$
$$y(t+i|t) = x(t+i-1|t) + u(t+i-1|t) \tag{9}$$
$$v_{xmin} < v_x < v_{xmax} \tag{10}$$
$$u_{min} < u(t+i-1|t) < u_{max} \tag{11}$$
$$\Delta u_{min} < u(t+i-1|t) - u(t+i-2|t) < \Delta u_{max} \tag{12}$$

where $(t + i/t)$ index indicate the values at future time $t + i$ and predicted at current time t. N is the prediction horizon. The vector of slack variables at time t is denoted by s_i.The tracking quadratic term, changes in inputs, inputs, potential field functions, and slack variables compose the objective function. The predicted potential field, as well as quadratic terms of tracking, inputs, changes in inputs, and slack variables, are all included in the objective function, with weighting matrices Q, R, S, and P, respectively. The predicted states are obtained by (8). The tracking output is calculated by (9). The constraints of speed and octagon approximation are applied as soft constraints represented in Eqs. (10). To satisfy the limitations of actuator, the inputs of control and their changes are constrained in (11) and (12) where u_{min} and u_{max} are the lower and upper bounds matrices of control input, and Δu_{min} and Δu_{max} are the lower and upper bounds matrices of the control inputs changes.

3. Results and discussion

3.1 Test scenario

The most challenging problems in the field of autonomous vehicles are path planning and control design. In structured and dynamic environments such as roads, route planning consists of both global and local path planning, with global path planning being utilized in conjunction with local path planning. Global path planning is a lengthy and deliberate procedure that is handled to develop long-distance routes to a destination. Local path planning, on the other hand, is a quicker procedure that is utilized for short-distance pathways and deals with duties like obstacle avoidance, comfort, safety, and vehicle stability. This planner is more responsive, as it operates in real-time.

Driving on organized roadways may be broken down into two fundamental vehicle maneuvers: lane keeping and lane switching. The primary goal of lane-keeping is to follow a vehicle and maintain its present position by changing its direction and distance from the lane center on a continual basis. Overtaking, obstacle avoidance and road departure are the most prevalent reasons for a vehicle to shift its present lane. The move may change depending on the route, lane, and obstacles on the road. In reality, there are several movements to be made. The performance of path planning systems may be assessed by watching these moves, as well as safety and road rules. If the vehicle's path is safe, the vehicle can stay in its lane. A lane change must be planned and implemented if this is not the case. This lane change occurs when the vehicle reaches the end of the road or encounters another barrier in its own route. If there are no obstacles or other vehicles on the targeted lane at the end of the road, lane switching is completed. Otherwise, the vehicle should slow down and perhaps stop before reaching the lane's end. When a vehicle encounters an obstacle in its path, it must predict the obstacle's path. The vehicle may pass the obstacle if there is adequate lateral distance; nevertheless, the vehicle is changing lanes to overtake. Otherwise, the vehicle should come to a complete stop in front of the obstacle or cross it. These are a few examples of movements that occur on the highway. Two scenarios are provided to evaluate the performance of autonomous vehicles: On both straight and curving roads, retain your lane. Maintaining a certain space between the ego vehicle and the vehicle in front of it.

3.2 Simulation

Since the provided potential field is a non-convex and nonlinear function, the optimization problem is non-convex and nonlinear, therefore solving it is expensive. Thus, the problem is converted into a quadratic and convex problem to reduce computing time. Convex functions are used to approximate PFs for this purpose. The obtained convex function is then approximated using the second-order Taylor series by a quadratic function. Around the nominal point, the resulting function is a near convex quadratic approximation of the original function. The resulting gradient is the same as the original function's gradient. The approximated function's Hessian matrix is the Frobenius norm's nearest positive definite matrix to the original function's Hessian matrix. Although the quadric approximation of the PFs increases the computation time, it is insignificant compared to the time required to solve a nonlinear optimization problem [29].

The problem of optimal control is a convex quadratic optimization problem when these PFs are used. This problem is related to a nonlinear problem that may be solved in one step using Sequential Quadratic Programming (SQP). Boggs et al. [30] calculate an upper bound for the optimization error of each SQP sequence, where this error is the difference between the sequence outcome and the local minimum of the nonlinear problem about the problem's initial value. According to this upper bound, if the problem's initial value is closer to a minimum, the optimization error will be reduced. The predicted vehicle position will be equal to the vehicle's location at a minimum point. Furthermore, in the Hessian matrix, the closer estimated PFs are to their minimum values, the lower the optimization error. As a result, a PF with a lower convex quadratic approximation error and a lower variation of the Hessian matrix about the problem's initial value has a lower optimization error.

In the following, the suggested MPC's performance on an autonomous vehicle is simulated and tested in terms of maneuverability, road regulation, and obstacle avoidance. The YALMIP toolbox in MATLAB/Simulink and the fmincon solver SQP are used to solve the MPC formulation. The features of a controller for a dry road are shown in **Table 1**. The vehicle is moving at 20 m/s, and the controller time step is 50 milliseconds.

Figure 5(a) depicts the first scenario, which is relevant to path planning on a regular highway. Based on offline lane marking and mapping waypoints, a 4th order polynomial is used to approximate the road geometry. It's a two-lane, one-way road. The ego vehicle is shown as a dashed blue circle traveling on lane 1, while the obstacle vehicle is shown as a red circle moving on the other lane. This scenario's

Parameter	Value	Unit
m	1625	*kg*
I_z	2865.6	$kg.m^2$
l_f	1.108	*m*
l_r	1.502	*m*
N	10	—

Table 1.
The controller parameters.

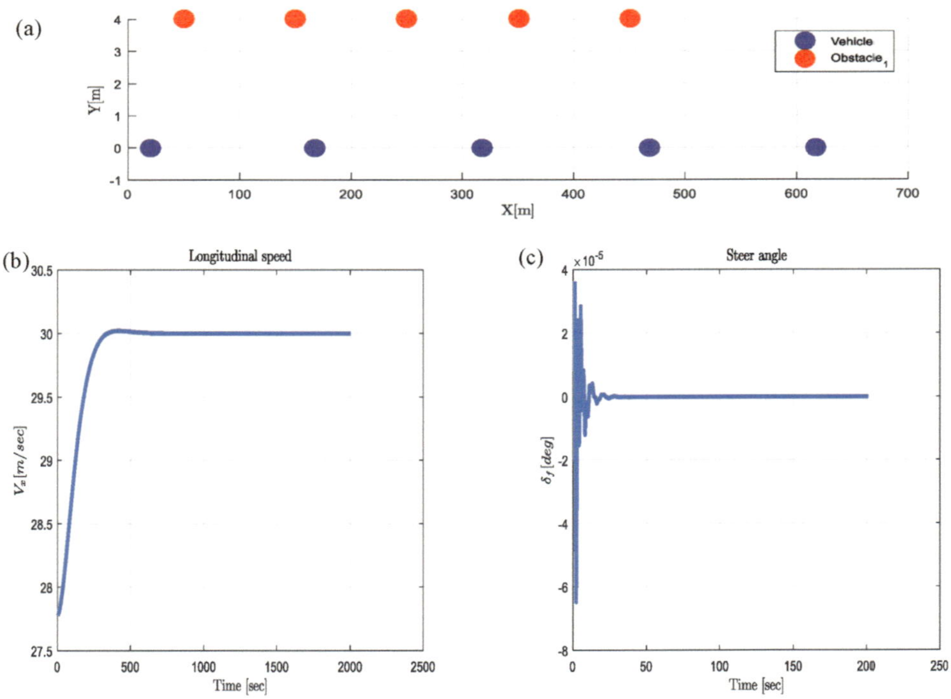

Figure 5.
(a) The path planning in the straight road with keeping lane, (b) longitudinal speed of the vehicle, (c) steering angle for the first scenario.

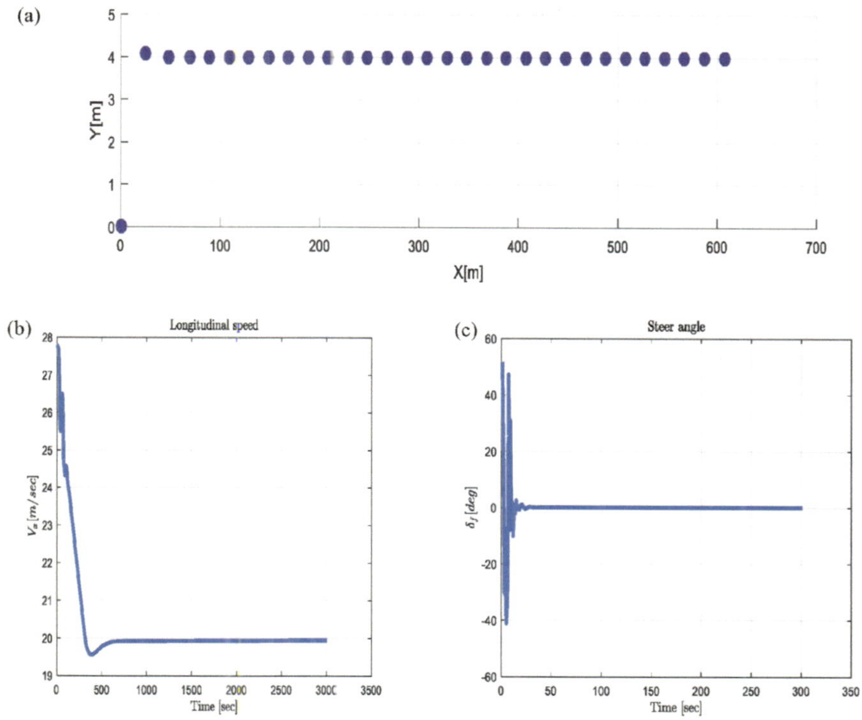

Figure 6.
(a)Changing lane the vehicle in the straight road and decrease decreases the speed from 27 m/sec to 20 m/sec, (b) longitudinal speed, (c) steering angle and lateral acceleration for the second scenario.

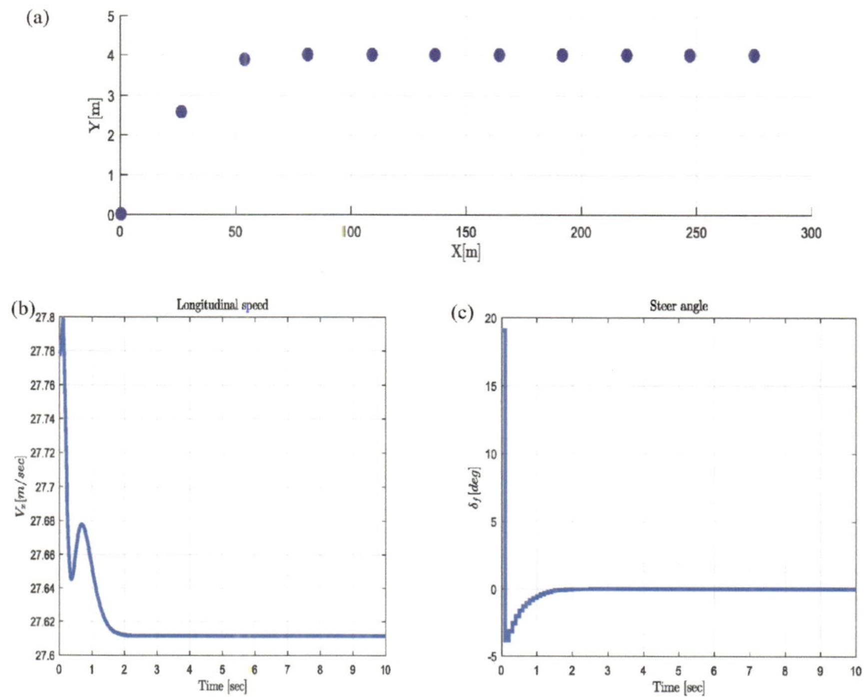

Figure 7.
(a) Changing lane the vehicle in the straight road, (b) longitudinal speed, (c) steering angle and lateral acceleration for the third scenario.

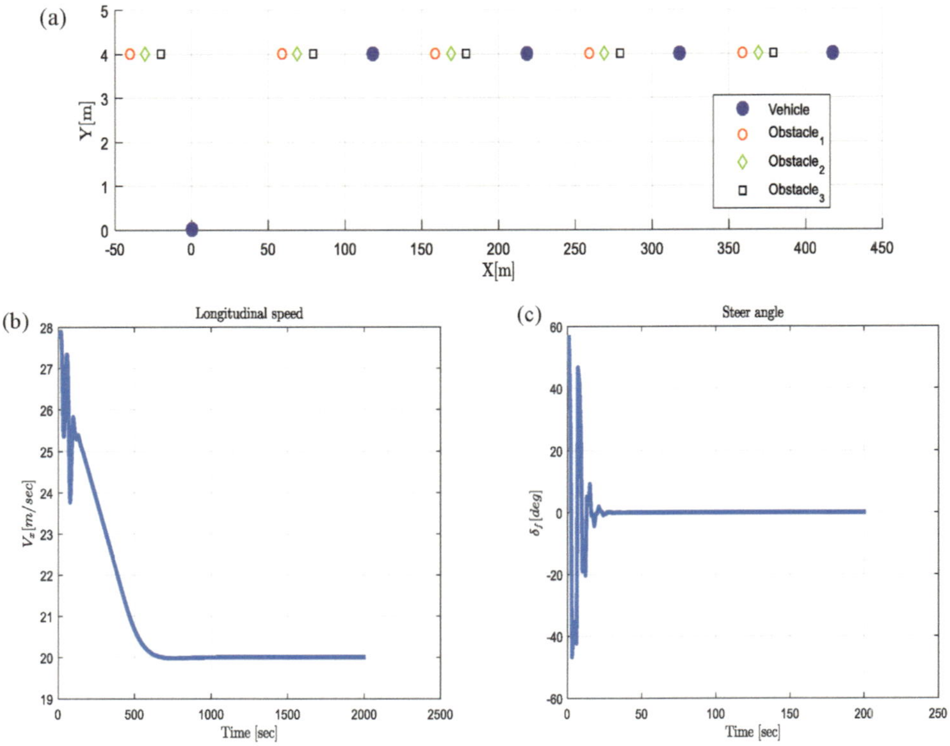

Figure 8.
(a) Changing lane along with three obstacles in the straight road, (b) longitudinal speed, (c) steering angle for the last scenario.

main goal is to demonstrate lane-keeping ability on a straight road. The path of the ego and the obstacle vehicle is represented by a sequence of circles. The ego vehicle's desired speed is higher than the obstacle vehicle in the other lane. The obstacle is moving at a speed of 20 m/sec, vehicle tries to increase its speed to 30 m/sec without changing the lane. **Figure 5(b)** depicts the longitudinal speed of the vehicle. **Figure 5(c)** depicts the steering angle of the vehicle.

In the second scenario, the ego vehicle starts on lane 1. The vehicle just tries to change the lane and decreases the speed from 27 m/sec to 20 m/sec and changes its front wheel angle (**Figures 6**).

In the next scenario, the ego vehicle starts on lane 1 and tries to change the lane, however, it remains its speed at 27 m/sec and changes its front wheel angle (**Figure** 7).

Figure 8 shows a merging maneuver along with three obstacles which is moving on the lane 2 and the ego is moving on lane 1. It should change its lane from the lane 1 to the lane 2 while avoiding a possible collision to the obstacles. Due to the lack of lateral distance between the vehicle and obstacle, the vehicle cannot immediately and safely merge between them. In this scenario, the potential field used for obstacle keeps the vehicle away from the lane 2 when the obstacle is passing from another lane. Additionally, the potential field is used for a static obstacle to avoid collision. Obstacles are moving with speed 20 m/sec, vehicle tries to come in their lane with decreasing its speed and the front wheel angle. Therefore, the vehicle reduces its speed and sometimes stops before reaching to the vehicle and after passing obstacle, the vehicle changes safely its lane and then the road and lane centering potential fields keep the vehicle for going out from the road.

4. Conclusion

In this paper, we present a path planning method for autonomous vehicles in dynamic settings that is based on model predictive controllers. We offer two alternative possible field functions for the road, center of the lane, and barriers or surrounding vehicles to avoid collision and assure vehicle safety. By definition, the problem is a nonlinear optimal control problem. To formulate the problem of path planning using a quadratic objective function, the MPC framework is chosen. The computing load is considerably reduced when the problem is approximated as a convex quadratic problem. This function makes it easier to leverage the vehicle dynamics and system limitations inside the MPC framework to calculate lane keeping and lane shifting maneuvers. Simulations are used to compare the computing time and performance of nonlinear and quadratic issues. The results demonstrate that the quadratic formulation outperforms the nonlinear variant in terms of performance. Several simulations are run to analyze various possibilities. The findings show that the suggested path planning algorithm can generate safe and pleasant pathways for self-driving vehicles. Because the vehicle dynamics are utilized as the predicted model, the planned path is an ideal path based on the vehicle dynamics.

Author details

Zahra Elmi[1*] and Soheila Elmi[2]

1 Department of Computer Engineering, Istanbul Sabahattin Zaim University, Istanbul, Turkey

2 Department of Electronics Engineering, Faculty of Electrical and Computer Engineering, University of Tabriz, Tabriz, Iran

*Address all correspondence to: zahra.elmi@izu.edu.tr

References

[1] Ravankar AA, Ravankar A, Emaru T, Kobayashi Y. Hpprm: Hybrid potential based probabilistic roadmap algorithm for improved dynamic path planning of mobile robots. IEEE Access. 2020 Dec 8; 8:221743-221766.

[2] Lin P, Choi WY, Chung CC. Local Path Planning Using Artificial Potential Field for Waypoint Tracking with Collision Avoidance. In2020 IEEE 23rd International Conference on Intelligent Transportation Systems (ITSC) 2020 Sep 20 (pp. 1-7). IEEE.

[3] Kyaw PT, Paing A, Thu TT, Mohan RE, Le AV, Veerajagadheswar P. Coverage path planning for decomposition reconfigurable grid-maps using deep reinforcement learning based travelling salesman problem. IEEE Access. 2020 Dec 15;8: 225945-225956.

[4] Chen G, Luo N, Liu D, Zhao Z, Liang C. Path planning for manipulators based on an improved probabilistic roadmap method. Robotics and Computer-Integrated Manufacturing. 2021 Dec 1;72:102196.

[5] Huo L, Zhu J, Wu G, Li Z. A novel simulated annealing based strategy for balanced UAV task assignment and path planning. Sensors. 2020 Jan;20(17): 4769.

[6] Wang L, Kan J, Guo J, Wang C. 3D path planning for the ground robot with improved ant colony optimization. Sensors. 2019 Jan;19(4):815.

[7] Shao S, Peng Y, He C, Du Y. Efficient path planning for UAV formation via comprehensively improved particle swarm optimization. ISA transactions. 2020 Feb 1;97:415-430.

[8] Elmi Z, Efe MÖ. Online path planning of mobile robot using grasshopper algorithm in a dynamic and unknown environment. Journal of Experimental & Theoretical Artificial Intelligence. 2021 May 4;33(3):467-485.

[9] Sipahioglu A, Yazici A, Parlaktuna O, Gurel U. Real-time tour construction for a mobile robot in a dynamic environment. Robotics and Autonomous Systems. 2008 Apr 30;56(4):289-295.

[10] Kim YH, Park JB, Son WS, Yoon TS. Modified turn algorithm for motion planning based on clothoid curve. Electronics Letters. 2017 Dec 7;53(24): 1574-1576.

[11] Yang S, Wang Z, Zhang H. Kinematic model based real-time path planning method with guide line for autonomous vehicle. In2017 36th Chinese Control Conference (CCC) 2017 Jul 26 (pp. 990-994). IEEE.

[12] Sudhakara P, Ganapathy V, Sundaran K. Optimal trajectory planning based on bidirectional spline-RRT* for wheeled mobile robot. In2017 Third International Conference on Sensing, Signal Processing and Security (ICSSS) 2017 May 4 (pp. 65-68). IEEE.

[13] Miao H, Tian YC. Dynamic robot path planning using an enhanced simulated annealing approach. Applied Mathematics and Computation. 2013 Oct 1;222:420-437.

[14] Chaudhary KL, Ghose D. Path planning in dynamic environments with deforming obstacles using collision cones. In2017 Indian Control Conference (ICC) 2017 Jan 4 (pp. 87-92). IEEE.

[15] D'Arcy M, Fazli P, Simon D. Safe navigation in dynamic, unknown, continuous, and cluttered environments. In2017 IEEE International Symposium on Safety, Security and Rescue Robotics (SSRR) 2017 Oct 11 (pp. 238-244). IEEE.

[16] Karaman S, Frazzoli E. Incremental sampling-based algorithms for optimal motion planning. Robotics Science and Systems VI. 2010 May 5;104(2).

[17] Kamil F, Hong TS, Khaksar W, Moghrabiah MY, Zulkifli N, Ahmad SA. New robot navigation algorithm for arbitrary unknown dynamic environments based on future prediction and priority behavior. Expert Systems with Applications. 2017 Nov 15; 86:274-291.

[18] Zhang K, Sprinkle J, Sanfelice RG. A hybrid model predictive controller for path planning and path following. InProceedings of the ACM/IEEE Sixth International Conference on Cyber-Physical Systems 2015 Apr 14 (pp. 139-148).

[19] Shen C, Shi Y, Buckham B. Model predictive control for an AUV with dynamic path planning. In2015 54th Annual Conference of the Society of Instrument and Control Engineers of Japan (SICE) 2015 Jul 28 (pp. 475-480). IEEE.

[20] Liu C, Lee S, Varnhagen S, Tseng HE. Path planning for autonomous vehicles using model predictive control. In2017 IEEE Intelligent Vehicles Symposium (IV) 2017 Jun 11 (pp. 174-179). IEEE.

[21] Gray A, Gao Y, Lin T, Hedrick JK, Tseng HE, Borrelli F. Predictive control for agile semi-autonomous ground vehicles using motion primitives. In2012 American Control Conference (ACC) 2012 Jun 27 (pp. 4239-4244). IEEE.

[22] Carvalho A, Gao Y, Gray A, Tseng HE, Borrelli F. Predictive control of an autonomous ground vehicle using an iterative linearization approach. In16th International IEEE conference on intelligent transportation systems (ITSC 2013) 2013 Oct 6 (pp. 2335-2340). IEEE.

[23] Ji J, Khajepour A, Melek WW, Huang Y. Path planning and tracking for vehicle collision avoidance based on model predictive control with multiconstraints. IEEE Transactions on Vehicular Technology. 2016 Apr 22;66 (2):952-964.

[24] Worthmann K, Mehrez MW, Zanon M, Mann GK, Gosine RG, Diehl M. Model predictive control of nonholonomic mobile robots without stabilizing constraints and costs. IEEE transactions on control systems technology. 2015 Oct 29;24(4):1394-1406.

[25] Gao Y, Gray A, Tseng HE, Borrelli F. A tube-based robust nonlinear predictive control approach to semiautonomous ground vehicles. Vehicle System Dynamics. 2014 Jun 3;52 (6):802-823.

[26] Brown M, Funke J, Erlien S, Gerdes JC. Safe driving envelopes for path tracking in autonomous vehicles. Control Engineering Practice. 2017 Apr 1;61:307-316.

[27] Silva-Ortigoza R, Márquez-Sánchez C, Carrizosa-Corral F, Hernández-Guzmán VM, Garcia-Sánchez JR, Taud H, Marciano-Melchor M, Alvarez-Cedillo JA. Obstacle Avoidance Task for a Wheeled Mobile Robot: A Matlab-Simulink-Based Didactic Application. MATLAB: Applications for the Practical Engineer. 2014 Sep 8:79-102.

[28] Elmi Z, Efe MÖ. Path planning using model predictive controller based on potential field for autonomous vehicles. InIECON 2018-44th Annual Conference of the IEEE Industrial Electronics Society 2018 Oct 21 (pp. 2613-2618). IEEE.

[29] Tanaka M, Nakata K. Positive definite matrix approximation with condition number constraint. Optimization Letters. 2014 Mar;8(3):939-947.

[30] Boggs PT, Tolle JW. Sequential quadratic programming. Acta numerica. 1995 Jan;4:1-51.

Chapter 6

Rolling Biped Polynomial Motion Planning

Santiago de J. Favela Ortíz and Edgar A. Martínez García

Abstract

This work discloses a kinematic control model to describe the geometry of motion of a two-wheeled biped's limbs. Limb structure is based on a four-bar linkage useful to alleviate damping motion during self-balance. The robot self-balancing kinematics geometry combines with user-customized polynomial vector fields. The vector fields generate safe reference trajectories. Further, the robot is forced to track the reference path by a model-based time-variant recursive controller. The proposed formulation showed effectiveness and reliable performance through numerical simulations.

Keywords: rolling-biped, path planning, motion control, navigation, underactuation

1. Introduction

Motion planning is an essential function and a critical aspect in robotics engineering. Motion planning allows increasing the robot's degree of autonomy. Fundamentally a wide area of robotic tasks needs planning models such as: transportation and vehicular technology, service robotics, search, exploration, surveillance, biomedical robotic applications, spatial deployment, industry, and so forth. Moreover, motion planning is inherently impacted by the degree of holonomy and kinematic constraints in all robotic modalities: robot arms, legged robots, rolling platforms, marine/underwater vehicles, aerial robots, and including their end effectors. Depending on the robotic application, motion planning is designed either global or local. When the robot has an environmental map in advance, it is called global planning even with possibility to globally optimize routes. Alternatively, when there is only robot's local sensor data and the whole environment is unknown, it uses feedback from local observations. Planning methods can be generalized into four types: deterministic (based on mathematical numeric/analytic functions and models) [1–3], stochastic (recursive numerical methods based on probabilistic uncertainties) [4–6], heuristic (algorithms based on logical control and human-heuristic decision-making) [7–10], and mixed planning methods [11–13].

In this chapter, a kinematic motion/path planning method for path tracking of an inverted pendulum self-balancing rolling biped is deduced and discussed. This work is focused on the rolling biped motion modeling and simulation of the robot shown in **Figure 1**. The principal component of a rolling biped is self-balancing by controlling its pitch motion through in-wheel motors that allow rolling motion (inverted-pendulum-like). The robot's yaw motion is accomplished by the angular

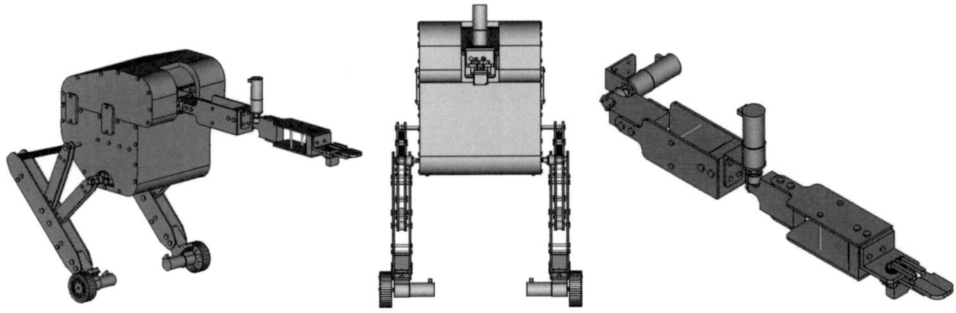

Figure 1.
Rolling biped platform views. (a) Isometric view. (b) Front-view. (c) Onboard robotic arm.

velocity resulting from the differential lateral speeds, which is a nonholonomic constrained model. The robot's design is purposed for missions to collect solid garbage in outdoors (a park), similar to other works [14]. Nevertheless, the robotic mission/task is out of this chapter's scope, instead a detailed geometry of motion is described in three parts: (i) motion planning for biped's balance, (ii) navigational path generation, and (iii) path tracking control. The work [15] proposed a balancing and dynamic speed control of a unicycle robot based on variable structure and linear quadratic regulator to follow a desired trajectory. The work [16] modeled a wheeled bipedal robot with analytic solutions of closed-form expressions in kinematic control loops. The work [17] reported a self-balancing two-wheeled robot with a manipulator on-board, using auto-balancing system to maintain force equilibrium. The work [18] applied a proportional integral derivative (PID) control and active disturbance rejection control to balance and steer a two-wheeled self-balancing robot modeled by Lagrange formula. In [19], an adaptive robust control of a self-balancing two-wheeled underactuated robot to estimate uncertainty bound information, using deterministic system performance by Lyapunov method, was reported. In [20], a navigational two-wheeled self-balancing robot control using a PD-PI controller based on the Kalman filter algorithm was reported. Similarly, [21] used variable structure combining proportional integral differential controllers for balance and locomotion deriving a kinematic model based on the center of gravity constraint. Lagrangian-based with Kane's approach for dynamic balancing was reported in [22]. Moreover, [23] conducted a study using model predictive control for trajectory tracking of an inverted-pendulum wheeled robot. The work [24] reported a self-balancing robot controller using Euler-Lagrange and geometric control, and planar motion is controlled by logarithmic feedback and Lie group exponential coordinates. The work [25] developed a balancing and trajectory tracking system for an inverted-pendulum wheeled robot using a Lagrange-based backstepping structure variable method. This work's main contributions are an original design of limbs based on four-bar linkages to alleviate damping motion yielded from irregular terrains, from which a kinematic balancing condition is deduced. Further, polynomial vector fields with limit conditions are deduced from user-customized interpolation functions as path-generator models to yield safe routes in advance. Moreover, a recursive time-varying kinematic controller forces the robot to track resulting routes. The proposed system is demonstrated at the level of simulation. This chapter is organized in the following sections. Section 2 deduces the limb kinematics and its effects in the biped's balance. Section 3 presents the polynomial approach to trajectory generation by directional fields. Section 4 describes a navigation recursive controller for path tracking control. Finally, Section 5 presents the work's conclusions.

2. Balancing motion planning model

The main issue of an inverted-pendulum-like rolling biped is its capability to self-balance by controlling its pitch angle through the wheels velocity longitudinally (**Figure 2(b)**). This section deduces the balancing kinematics of the limb's planar linkage shown in **Figure 2(a)**. As a difference from other biomechanical inspired muscle-tendon limbs [26], in this work each limb is comprised of a four-bar linkage operating as a double crank, where bars r and d are linked by a coupling link l with limited rotary angles.

The expressions provided in Proposition 2.1 are obtained from deductions in Appendix A. The passive joint point $\mathbf{P}_1 = (x_1, y_1)^\top$ mutually depends on the analytic model of the joint located at $\mathbf{P}_2 = (x_2, y_2)^\top$ to describe a rotary planar motion, proposed by

Proposition 2.1 Limbs motion estimation. *The four-bar linkage's passive joint* $\mathbf{P}_2$ *is analytically described by.*

$$\mathbf{P}_2 = d\begin{pmatrix} \cos(\theta) \\ \sin(\theta) \end{pmatrix} + \begin{pmatrix} \Delta x \\ \Delta y \end{pmatrix}. \tag{1}$$

Therefore, based on Definitions 4.1–4.3 of Appendix A, the model solution $\mathbf{P}_2$ *is*

$$\begin{pmatrix} x_2 \\ y_2 \end{pmatrix} = d\begin{pmatrix} \cos\left(2\arctan\left(\dfrac{-Q \pm \sqrt{Q^2 - 4PR}}{2P}\right)\right) \\ \sin\left(2\arctan\left(\dfrac{-Q \pm \sqrt{Q^2 - 4PR}}{2P}\right)\right) \end{pmatrix} + \begin{pmatrix} \Delta x \\ \Delta y \end{pmatrix} \tag{2}$$

From Eq. (56) deduced in Appendix A, there are two possible solutions for θ, described as opened or crossed motion. For this work, the type of motion should be $\sqrt{Q^2 - 4PR} \forall Q^2 \leq 4PR$ that must be satisfied. Therefore, by using expressions (47), (52)–(56), the plots shown in 3 are obtained.

Furthermore, the wheels position is tracked by $\mathbf{P}_f$ (**Figure 2(a)**). This design assumes the robot's center of gravity (cog) located at the same length $f \equiv a$

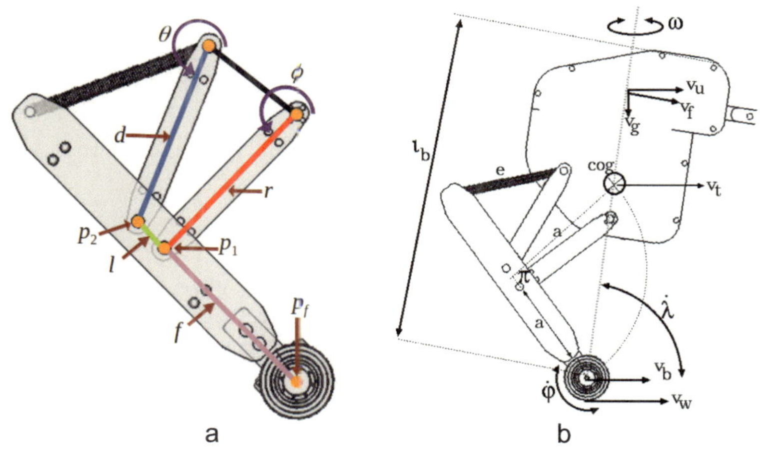

Figure 2.
Rolling biped kinematic constrains. (a) Limb's planar linkage parameters. (b) Balancing parameters.

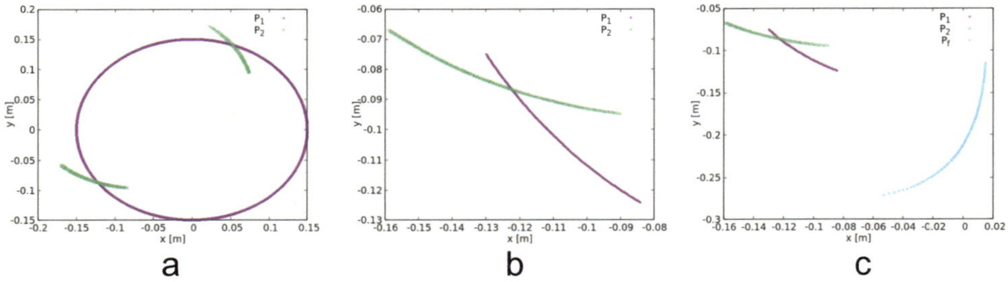

Figure 3.
Limb motion planning produced with parameters r = 0.15 *m*, d = 0.15 *m*, l = 0.03 *m*, Δx = −0.07 *m*, Δy = 0.05387 *m*, ∀201° ≤ *ϕ* ≤ 236°.

(both in **Figure 3**) that is projected within the biped's body. Therefore, the cog is described by Definition 2.1:

Definition 2.1 Robot's center of gravity (cog). *The robot's cog is assumed to be located at Cartesian body's position,*

$$x_{cog} = \left(\frac{x_2 - x_1}{2} + a\right) \cos\left(\pi + \arctan\left(\frac{y_2 - y_1}{x_2 - x_1}\right)\right) \tag{3}$$

and

$$y_{cog} = \left(\frac{y_2 - y_1}{2} + a\right) \sin\left(\pi + \arctan\left(\frac{y_2 - y_1}{x_2 - x_1}\right)\right). \tag{4}$$

Hence, in accordance to Definition 2.1, it follows that the robot's falling speed v_f due to vertical unbalance is

$$v_f = \sqrt[2]{\left(\frac{dx_{cog}}{dt}\right)^2 + \left(\frac{dy_{cog}}{dt}\right)^2}. \tag{5}$$

Taking into account that the robot is a dual differential drive kinematic structure, where v_R and v_L are the right-sided and left-sided velocities, respectively

$$\frac{ds}{dt} = \frac{r}{2}\left(\frac{d\varphi_R}{dt} + \frac{d\varphi_L}{dt}\right), \tag{6}$$

where s is the robot displacement over the ground, and r is the wheel's radius [m]. Likewise, the dual differential velocity is

$$v_{diff} = \frac{d\varphi_R}{dt} - \frac{d\varphi_L}{dt}. \tag{7}$$

Hence, the robot's angular velocity ω [rad/s] in terms of its differential speed v_{diff} and constrained by its wheels lateral baseline distance b_l and v_{diff} is substituted to yield:

$$\omega = \frac{2v_{diff}}{b_l} = \frac{2r_w}{b_l}\left(\frac{d\varphi_R}{dt} - \frac{d\varphi_L}{dt}\right). \tag{8}$$

The biped's falling angle and angular falling speed are λ [rad] and $\dot{\lambda}$ [rad/s], respectively. Let v_f [m/s] be the falling velocity affecting the robot's balance

$$v_f = \ell_b \dot{\lambda}_t, \tag{9}$$

which makes the robot's pitch turn around the wheel's center. Unbalancing velocity v_u [m/s] is an horizontal component such that

$$v_u = v_f \cos\left(\frac{\pi}{2} - \lambda\right) = \ell_b \dot{\lambda} \cos\left(\frac{\pi}{2} - \lambda\right). \tag{10}$$

The wheel's axis point moves at balancing speed v_b that is parallel to the ground. Wheel's tangential speed is $v_w = r\dot{\varphi}$, and linear motion speed is the balancing velocity underneath the biped's body at the wheel's axis,

$$v_b = \frac{v_w}{2} = \frac{r\dot{\varphi}_w}{2} \tag{11}$$

Therefore, the balancing condition is described by Definition 2.2:

Definition 2.2 Kinematic balancing condition. *The robot's vertical equilibrium condition is assumed for balancing and unbalancing velocities of equal magnitudes (or very approximated). Thus,*

$$v_u - v_b = 0, \tag{12}$$

and redefining, let $\dot{\varphi}_B$ be the wheel's balancing angular speed,

$$\ell_b \dot{\lambda} \cos\left(\frac{\pi}{2} - \lambda\right) - \frac{r_w \dot{\varphi}_B}{2} = 0. \tag{13}$$

Therefore, it is of interest to find the wheel's rotary speed that balances the biped motion and from Definition 2.2, the following Proposition 2.2 arises,

Proposition 2.2 Robot's balancing speed. *The wheel's rolling velocity to satisfy the robot's body balance is proposed as*

$$\dot{\varphi}_B = \frac{2\ell_b}{r_w} \dot{\lambda}_t \cos\left(\frac{\pi}{2} - \lambda\right). \tag{14}$$

and if and only if $\dot{\varphi}_R = \dot{\varphi}_L$ must exist, assuming that in such period of time, the robot's yaw $\omega_t \approxeq 0$ allowing longitudinal balancing equilibrium. Thus,

$$v_t = \begin{cases} r\dot{\varphi}_B, & \dot{\varphi}_R \approx \dot{\varphi}_L, \quad v_u - v_b \neq 0 \\ \frac{r}{2}(\dot{\varphi}_R + \dot{\varphi}_L), & v_u - v_b \approx 0 \end{cases} \tag{15}$$

Thus, let us redefine the state variables as $x_1 = \lambda$, $x_2 = \dot{x}_1$, $\dot{x}_2 = x_2^2 \tan\left(\frac{\pi}{2} - x_1\right)$. For an stability analysis for the balancing case when $v_t = r_w \dot{\varphi}_B$, and equilibrium condition $\ddot{\varphi}_B = 0$ occurs,

$$\frac{2\ell_b}{2}\left(\dot{x}_2 \cos\left(\frac{\pi}{2} - x_1\right) + \dot{x}_2{}^2 \sin\left(\frac{\pi}{2} - x_1\right)\right) = 0 \tag{16}$$

and dropping off the highest-order derivative of the system

$$\ddot{\lambda} = x_2^2 \tan\left(\frac{\pi}{2} - x_1\right) \tag{17}$$

The system total energy $E_k + E_p$ (kinetic plus potential) is a positive function, which is used as a Lyapunov candidate function, where the robot's translation motion $r_w \dot{\varphi}_B$ for equilibrium is

$$E_T = E_k + E_p = \frac{1}{2} m r_w^2 \dot{\varphi}_B^2 + mg\frac{\ell_b}{2}(1 - \cos(\lambda)) \quad (18)$$

and substituting $\dot{\varphi}_B$,

$$E_T = 2m\ell_b^2 \dot{\lambda}^2 \cos\left(\frac{\pi}{2} - \lambda\right) + mg\frac{\ell_b}{2}(1 - \cos(\lambda)). \quad (19)$$

For $V(x) = E_T$ and finding out that v(x) and $\dot{V}(x)$ fulfill $\dot{V}(0) = 0$. In addition, assuming that $-2\pi \ll \lambda \ll 2\pi$, then $\dot{V}(0) > \in D - 0$. Assuming that $V : D \rightarrow \mathbb{R}$ is a continuous differentiable function. Finally, the following Lyapunov criterion is satisfied (20),

$$\left(\frac{\partial V(x)}{\partial x_1}, \frac{\partial V(x)}{\partial x_2}\right) \cdot (x_2, \dot{x}_2)^{\top} = 0 \quad (20)$$

3. Polynomial vector fields

This section introduces the proposed path planning strategy to dynamically generate local paths. Two polynomial models are enhanced as vector fields to exert attractive and repulsive robot's accelerations. In principle, the desired acceleration functions are designed from the interpolating attractive/repulsive accelerations with respect to (w.r.t.) distance. The method used to fit points is the Lagrange formula:

$$a(\delta) = \sum_i \left(\prod_j \frac{\delta - \delta_i}{\delta_i - \delta_j}\right) a_i, \quad (21)$$

where the human-user establishes a desired numerical acceleration a_i w.r.t. a desired distance δ_i, from either goal or obstacle. Therefore, the following general cubic polynomial forms attractive a_A and repulsive a_R are obtained with their respective numeric coefficients λ_i and β_i,

$$a_A(\delta) = \lambda_0 + \lambda_1\delta + \lambda_2\delta^2 + \lambda_3\delta^3, \quad a_R(\delta) = \beta_0 + \beta_1\delta + \beta_2\delta^2 + \beta_3\delta^3 \quad (22)$$

It follows that, Definition 3.1 establishes the acceleration path planning model toward a goal of interest.

Definition 3.1 Planning toward a goal. *Given the kinematic parameters* $\lambda_0 = 0$, $\lambda_1 = \frac{18}{75}[s^{-2}]$, $\lambda_2 = \frac{18}{75}[m^{-1}s^{-2}]$, $\lambda_3 = \frac{3}{75}[m^{-2}s^{-2}]$, *the attraction path model* $-\nabla_\delta \kappa f^A(\delta)$ *is defined by*

$$f^A(\delta_g) = -\nabla_\delta \frac{3}{75}\left(6\delta_g + 6\delta_g^2 - \delta_g^3\right). \quad (23)$$

Assuming that the distance between the goal and the robot $\delta_g = \sqrt[2]{x^2 + y^2}$, *such that* $x \doteq x_g - x_r$ *and* $y \doteq y_g - y_r$.

Similarly, Definition 3.2 establishes the acceleration path planning model that avoids obstacles zones.

Definition 3.2 Planning obstacles avoidance. *Given the kinematic parameters* $\beta_0 = 5$, $\beta_1 = \frac{109}{440}$, $\beta_2 = -\frac{18}{55}[m^{-1}s^{-2}]$, $\beta_3 = \frac{3}{110}[m^{-2}s^{-2}]$, *the avoidance path model* $-\nabla_\delta f^R(\delta)$ *is defined by*

$$f^R(\delta_o) = -\nabla_\delta \kappa_R \left(5 + \frac{109}{440}\delta_o - \frac{18}{55}\delta_o^2 + \frac{3}{11}\delta_o^3 \right), \tag{24}$$

Assuming that the distance between the obstacle and the robot $\delta_0 = \sqrt[2]{x^2+y^2}$, *such that* $x \doteq x_o - x_r$ *and* $y \doteq y_o - y_r$.

Figure 4(a) shows the robot's instantaneous acceleration toward a goal of interest. The attraction acceleration starts when the goal-robot distance $\delta_g < 9$m. The planner allows start from $f^A \approxeq 0$ to realistically provide speeds physically possible. **Figure 4(b)** shows the robot's instantaneous acceleration away from obstacles. The avoidance acceleration starts when the obstacle-robot distance $\delta_o < 8$m. This avoidance planner f^R is faster than f^A to increase confidence against obstacles.

Therefore, extending to two-dimension Cartesian space, let us deduce the following algebraic process for the goal-attraction planner f^A:

$$\frac{\partial f^A}{\partial x} = -\kappa_A \left(\frac{6x}{\sqrt[2]{x^2+y^2}} + 12x - \frac{3x}{2}\sqrt[2]{x^2+y^2} \right), \tag{25}$$

as well as

$$\frac{\partial f^A}{\partial y} = -\kappa_A \left(\frac{6y}{\sqrt[2]{x^2+y^2}} + 12y - \frac{3y}{2}\sqrt[2]{x^2+y^2} \right). \tag{26}$$

Similarly, for the robot's acceleration to avoid obstacles, let us develop the following f^R, by substituting the functional form of δ_o into the gradient function

$$f^R(x,y) = -\nabla \kappa_R \left(5 + \frac{109}{440}\sqrt[2]{x^2+y^2} - \frac{18}{55}(x^2+y^2) + \frac{3}{110}(x^2+y^2)^{3/2} \right), \tag{27}$$

subsequently by applying the gradient operator,

$$\frac{\partial f^R}{\partial x} = -\frac{109x}{440\sqrt[2]{x^2+y^2}} + \frac{18x}{55} - \frac{9x}{110}\sqrt[2]{x^2+y^2} \tag{28}$$

and

$$\frac{\partial f^R}{\partial y} = -\frac{109y}{440\sqrt[2]{x^2+y^2}} + \frac{18y}{55} - \frac{9y}{110}\sqrt[2]{x^2+y^2} \tag{30}$$

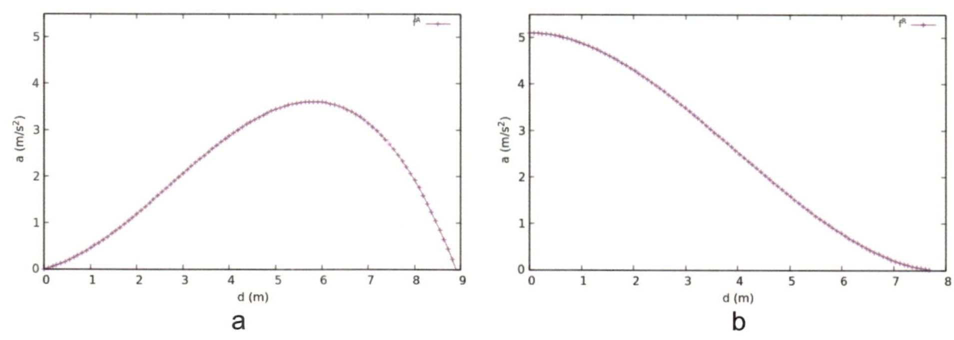

Figure 4.
Navigational robot's motion planning, desired acceleration w.r.t. distance. (a) Goal attractive motion. (b) Obstacle avoidance motion.

Thus, by substituting terms in both local planners, the general attractive motion planning is a negative function with amplitude coefficient $\kappa_A = \frac{3}{75}[s^{-2}]$ is represented by

$$\begin{pmatrix} \frac{\partial f^A}{\partial x} \\ \frac{\partial f^A}{\partial y} \end{pmatrix} = -\kappa_A \left(\frac{6}{\delta_t} + 12 - \frac{3}{2}\delta_g \right) \begin{pmatrix} x_g - x_t \\ y_g - y_t \end{pmatrix} \tag{31}$$

The attractive field is a positive function working within limits $0 \le \delta_g < 9$, thus

$$\lim_{\delta_g \to 9} f^R(\delta_g) = 0; \quad \text{and} \quad \lim_{\delta_g \to 6} f^R(\delta_g) = 3.75 \tag{32}$$

and the repulsive with $\kappa_R = \frac{1}{440}[s^{-2}]$

$$\begin{pmatrix} \frac{\partial f^R}{\partial x} \\ \frac{\partial f^R}{\partial y} \end{pmatrix} = -\kappa_R \left(\frac{109}{\delta_o} - 288 + 36\delta_o \right) \begin{pmatrix} x_o - x_t \\ y_o - y_t \end{pmatrix}. \tag{33}$$

The repulsive field is a negative function that works within range $0 \le \delta_o \le 8$,

$$\lim_{\delta_o \to 8} f^R(\delta_o) = 0; \quad \text{and} \quad \lim_{\delta_o \to 0} f^R(\delta_o) = 5. \tag{34}$$

The direction fields produced are shown in **Figure 5**, where for both cases the coordinates origin represents either goal or obstacle locations.

Moreover, when neither goals of interest nor obstacles are within the observation field of the robot, it must keep navigating along a prior route plan. The routing plan is map comprised of a sequence of Cartesian points $\mathbf{g}_k \in \mathbb{R}^2$, $\mathbf{g}_k = (x, y)^\top$.

When the robot accomplishes either f^A or f^R, it continues toward the following route point, expressed in terms of unit vectors:

$$\mathbf{a}_t^o = a^o \frac{\mathbf{g}_{k+1} - \mathbf{g}_k}{\|\mathbf{g}_{k+1} - \mathbf{g}_k\|}, \tag{35}$$

where a^o is an ideal or averaged acceleration toward the next route point $\mathbf{g}_2$ from $\mathbf{g}_1$. Therefore, the total path planner mode is given by Proposition 3.1.

Proposition 3.1 Total mission path planning. *The total path planning model subjected to adaptive environmental changes is proposed as a vector fields sum:*

$$\mathbf{a}_t = \mathbf{a}_t^o + \sum_g \left[\kappa_A \left(\frac{3\delta_g}{2} - \frac{6}{\delta_g} - 12 \right) \begin{pmatrix} x_g - x_t \\ y_g - y_t \end{pmatrix} + \sum_o \kappa_R \left(288 - \frac{109}{\delta_o} - 36\delta_o \right) \begin{pmatrix} x_0 - x_t \\ y_o - y_t \end{pmatrix} \right] \tag{36}$$

and the robot's instantaneous navigational total velocity $\mathbf{v}_T \in \mathbb{R}^2$, $\mathbf{v}_T = (x_T, \dot{y}_T)^\top$ is obtained by

$$\mathbf{v}_T = \int_{t_1}^{t_2} \mathbf{a}_t \mathrm{d}t. \tag{37}$$

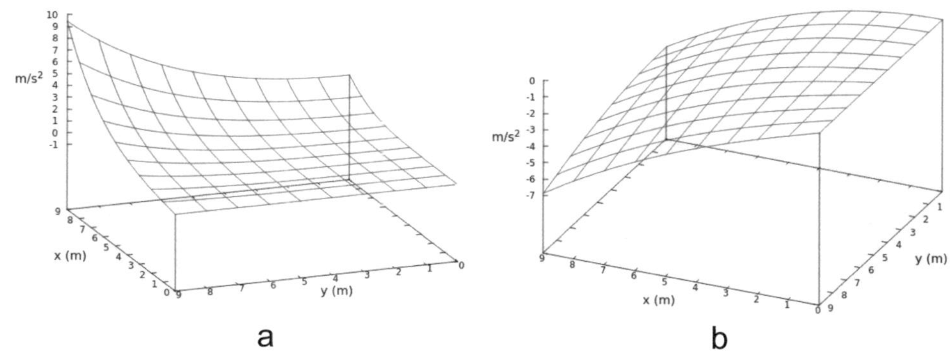

Figure 5.
Acceleration fields w.r.t. variations along the plane xy. *(a) Goals attractive motion generation. (b) Obstacles avoidance motion generation.*

Thus, to automatically limit the speed until the robot's maximal velocity $v_{\max}$, the real physical velocity v_{ph} is constrained by the final planning motion

$$v_{ph}\left(\frac{\|\mathbf{v}_T\|}{v_{max}}\right) = \begin{cases} \frac{\|\mathbf{v}_T\|}{v_{max}}, & \frac{\|\mathbf{v}_T\|}{v^{max}} < 1 \\ v_{max}, & \frac{\|\mathbf{v}_T\|}{v^{max}} \geq 1 \end{cases} \tag{38}$$

Figure 6 shows simulation results of the total navigation planner.

4. Path tracking

This section deduces an algebraic trajectory tracking linearized controller based on the kinematics geometry of the reference path: linear and angular velocities. Considering from previous Section 3, let v_{ph} be called the reference velocity that is going to be tracked. Likewise, let $(x_{1,2}, y_{1,2})$ be the next local planning coordinates to be reached (from location 1 to location 2). Hence, let us define $x_{1,2} \doteq x_2 - x_1$ and $y_{1,2} \doteq y_2 - y_1$. Such that $\dot{s}_{1,2}$ [m/s] is a segment of the planning speed. Hence, let us obtain the first-order derivative,

$$\dot{s}_{1,2} = \frac{d}{dt}\sqrt[2]{x_{1,2}^2 + y_{1,2}^2} = \frac{x_{1,2}\dot{x}_{1,2} + y_{1,2}\dot{y}_{1,2}}{\sqrt[2]{x_{1,2}^2 + y_{1,2}^2}}. \tag{39}$$

Similarly, by obtaining the desired robot's orientation $\theta_{1,2}$, its first-order derivative w.r.t. time is

$$\dot{\theta}_{1,2} = \frac{d}{dt}\left(\arctan\left(\frac{y_{1,2}}{x_{1,2}}\right)\right) = \frac{x_{1,2}\dot{y}_{1,2} - y_{1,2}\dot{x}_{1,2}}{x_{1,2}^2 + y_{1,2}^2}. \tag{40}$$

It follows that, the tracking control vector is $\dot{\mathbf{u}}_t \in \mathbb{R}^2$, such that $\dot{\mathbf{u}}_t = \left(\dot{s}_{1,2}, \dot{\theta}_{1,2}\right)^\top$. Therefore, by stating (39) and (40) as a system of linear equations, the first-order derivatives must simultaneously be solved:

$$\begin{aligned} \frac{x_{1,2}}{s_{1,2}}\dot{x}_{1,2} + \frac{y_{1,2}}{s_{1,2}}\dot{y}_{1,2} &= \dot{s} \\ -\frac{y_{1,2}}{s_{1,2}^2}\dot{x}_{1,2} + \frac{x_{1,2}}{s_{1,2}^2}\dot{y}_{1,2} &= \dot{\theta} \end{aligned}, \tag{41}$$

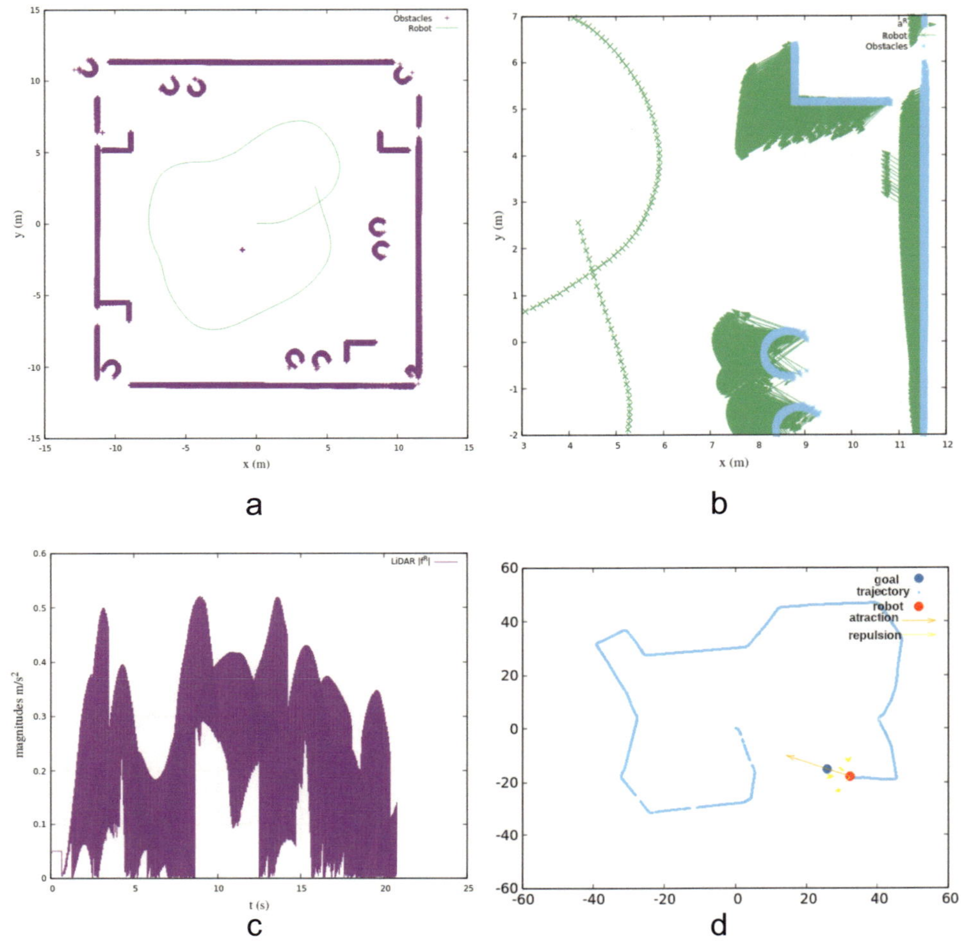

Figure 6.
Vector fields path generation. (a) High-density obstacles yielding robot's repulsion. (b) Obstacles repulsive field. (c) High density $\left\|\mathbf{f}^R\right\|$*. (d) Attractive goals along a route.*

thus, by factorizing both, the common term $1/s_{1,2}$ and the first-order derivatives, the matrix form of the forward tracking law is

$$\dot{\mathbf{u}} = \frac{1}{s_{1,2}} \begin{pmatrix} x_{1,2} & y_{1,2} \\ -\frac{y_{1,2}}{s_{1,2}} & \frac{x_{1,2}}{s_{1,2}} \end{pmatrix} \cdot \begin{pmatrix} \dot{x}_{1,2} \\ \dot{y}_{1,2} \end{pmatrix}. \tag{42}$$

Likewise, as the inverse tracking law is of our interest, it is inversely dropped off

$$\begin{pmatrix} \dot{x}_{1,2} \\ y_{1,2} \end{pmatrix} = s_{1,2} \begin{pmatrix} x_{1,2} & y_{1,2} \\ -\frac{y_{1,2}}{s_{1,2}} & \frac{x_{1,2}}{s_{1,2}} \end{pmatrix}^{-1} \cdot \dot{\mathbf{u}}. \tag{43}$$

For notation simplicity, let us redefine $\dot{\zeta}_t \doteq \left(\dot{x}_{1,2}, \dot{y}_{1,2}\right)^\top$ and the time-variant control matrix $\mathbf{K}_t$ as

$$\mathbf{K}_t = \begin{pmatrix} x_{1,2} & y_{1,2} \\ -\frac{y_{1,2}}{s_{1,2}} & \frac{x_{1,2}}{s_{1,2}} \end{pmatrix}, \tag{44}$$

Parameter	Value	Parameter	Value	Parameter	Value	Parameter	Value
l [m]	0.03	a [m]	0.25	$\kappa_A [s^2]$	3/75	I_x [kg²]	0.12
d [m]	0.15	b_l [m]	0.4	$\kappa_R [s^{-2}]$	1/440	I_y [kg²]	0.12
e^a [m]	0.08	ℓ_b [m]	0.75	a^o [m/s²]	0.5	I_z [kg²]	0.13
r [m]	0.15	m [kg]	0.76	r_w [m]	0.1	v_{max} [m/s]	1.5

[a]*High stiffness non-rigid bar.*

Table 1.
Simulation parameters deployed with high-density range scanner Hokuyo URG 04LX.

thus from Section 3, let $\boldsymbol{u}^{ref}$ be the global reference planning model during time segment $t_{1,2} \doteq t_2 - t_1$ such that,

$$\boldsymbol{u}^{ref} = \begin{pmatrix} v_{ph} \cdot t_{1,2} \\ \arctan\left(\frac{\dot{y}_T}{\dot{x}_T}\right) \end{pmatrix}. \tag{45}$$

Therefore, the recursive path tracking control law is stated as

$$\begin{aligned} 1: \quad \zeta_{t+1} &= \zeta_t + s_{1,2}\mathbf{K}_t^{-1} \cdot \left(\boldsymbol{u}^{ref} - \hat{\mathbf{u}}_t\right) \\ 2: \quad \mathbf{u}_{t+1} &= \mathbf{u}_t + \frac{1}{s_{1,2}}\mathbf{K}_t \cdot \left(\zeta_{t+1} - \hat{\zeta}_t\right) \end{aligned} \tag{46}$$

where $\hat{\mathbf{u}}_t$ is the instantaneous sensors observation of both components, displacement $\hat{s}$ and yaw $\hat{\theta}$. The prediction control speed vector ζ_{t+1} is the next desired local reference in line 2, while $\hat{\zeta}_t$ is the robot's Cartesian speed observer vector. Finally, $\mathbf{u}_{t+1}$ is the control vector global prediction (**Table 1**).

5. Conclusion

The mechanical design of the biped's lower-limb mechanical structure was configured as a double-crank four-bar linkage with passive-allowed motions. Motion planning began from determining the limbs' linkage positions causing the robot's height and pitch varying overtime producing unbalanced motions. Inferring balancing velocities to yield robot's vertical balance was possible and worked stable. The proposed balancing rolling condition was analyzed throughout its total energy model as a Lyapunov candidate function resulting stable in three criteria: $\mathbf{V}(0) = 0$, $v(x) = 0$, and $\dot{V}(x) = 0$. The proposed navigational approach allowed human-user to design short range-limited trajectories by cubic polynomials obtained from four empirical coordinates distance-acceleration by Lagrange interpolation. Both polynomial planning models have initial conditions $a_0(t_0) = 0[m/s^2]$. Thus, the robot did not require an infinite acceleration to reach a desired speed at $\mathbf{t}_0$. Physically, the model is applied to any motion in equilibrium. For goal attraction the maximal acceleration is reached at the 55% of the distance, subsequently decreases monotonically until the goal position.

The navigational general planning model is a set of partial derivatives model combined, allowing dynamic local planning among multiple obstacles, goals, and routes. The navigational general planning model worked as the reference model for the tracking control system. It consisted of a set of time-varying linear equations, with recursive feedback showing suitable performance. The work was implemented in simulation and coded in C++ under Linux. Future improvements will consider the inclusion of kinetic models and dynamic forces fitted to the proposed planer.

Appendix A

This appendix provides the algebraic deduction of the limb's Cartesian motion model described in Section 2.

Let $\mathbf{P}_1 = (x_1, y_1)^\top$ be the Cartesian point of passive joint between links b and ℓ, obtained by

$$\begin{pmatrix} x_1 \\ y_1 \end{pmatrix} = r \begin{pmatrix} \cos(\phi) \\ \sin(\phi) \end{pmatrix}. \tag{47}$$

Likewise, let $\mathbf{P}_2 = (x_2, y_2)^\top$ be the coordinate of passive joint between links r and ℓ, expressed by:

$$\begin{pmatrix} x_2 \\ y_2 \end{pmatrix} = r \begin{pmatrix} \cos(\phi) \\ \sin(\phi) \end{pmatrix} + \begin{pmatrix} \sqrt{l^2 - (y_1 - y_2)^2} \\ \sqrt{l^2 - (x_1 - x_2)^2} \end{pmatrix}. \tag{48}$$

Both expressions (47) and (48) are described in terms of ϕ, but can similarly be described in terms of the angle θ by

$$\begin{pmatrix} x_2 \\ y_2 \end{pmatrix} = d \begin{pmatrix} \cos(\theta) \\ \sin(\theta) \end{pmatrix} + \begin{pmatrix} \Delta x \\ \Delta y \end{pmatrix}, \tag{49}$$

where the distances Δx and Δy separate the joints *phi* and *theta* along the chassis, see **Figure 2(a)**. By algebraically treating the Cartesian components separately, let us equal the x components of (48) and (49) to form the equation

$$d\cos(\theta) + \Delta x = r\cos(\phi) + \sqrt{l^2 - (y_1 - y_2)^2}, \tag{50}$$

dropping off the squared root term of (50),

$$\sqrt{l^2 - (y_1 - y_2)^2} = d\cos(\theta) + \Delta x - r\cos(\phi) \tag{51}$$

and squaring both sides of the equality, the following is obtained,

$$l^2 - (y_1 - y_2)^2 = [d\cos(\theta) + \Delta x - r\cos(\phi)]^2 \tag{52}$$

and algebraically expanding the squared binomials and substituting y_1 and y_2 from expression (47) and (49) respectively into (52),

$$\begin{aligned} & d^2\cos^2(\theta) + 2\Delta x d\cos(\theta) - 2rd\cos(\theta)\cos(\phi) + (\Delta x)^2 \\ & \quad -2\Delta x r\cos(\phi) + r^2\cos^2(\phi) \\ & = l^2 - r^2\sin^2(\phi) + 2rd\sin(\theta)\sin(\phi) + 2\Delta y r\sin(\phi) \\ & \quad -d^2\sin^2(\theta) - 2\Delta y d\sin(\theta) - (\Delta y)^2 \end{aligned} \tag{53}$$

and organizing trigonometric terms by their degree, Eq. (53) becomes:

$$\begin{aligned}&2rd\cos(\theta)\cos(\phi)+2rd\sin(\theta)\sin(\phi)\\&\quad=d^2\cos^2(\theta)+d^2\sin^2(\theta)+(\Delta x)^2+(\Delta y)^2\\&\qquad+r^2\cos^2(\phi)+r^2\sin^2(\phi)+2\Delta xd\cos(\theta)-2\Delta xr\cos(\phi)\\&\qquad+2\Delta yd\sin(\theta)-2\Delta yr\sin(\phi).\end{aligned}\tag{54}$$

By factorizing the common terms of (47) and exchanging terms with suitable identities such as $\cos^2(\alpha)+\sin^2(\alpha)=1$ and $c^2=(\Delta x)^2+(\Delta y)^2$ in order to obtain Eq. (55)

$$\cos(\theta)\cos(\phi)+\sin(\theta)\sin(\phi)=\frac{r^2+d^2+c^2-l^2-2r[\Delta x\cos\phi+\Delta y\sin(\phi)]+2d\Delta x\cos(\theta)+2d\Delta y\sin(\theta)}{2rd}\tag{55}$$

In order to simplify the main expression, some constant terms are rewritten as

$$K_1=\frac{r^2+d^2+c^2-l^2}{2rd},\quad K_2-\frac{1}{d},\quad K_3=\frac{\Delta x}{r},\quad K_4=\frac{\Delta y}{r}.\tag{56}$$

Therefore, the following simplified algebraic expression is deduced:

$$\begin{aligned}&\cos(\theta)\cos(\phi)+\sin(\theta)\sin(\phi)=K_1+K_2[\Delta x\cos(\phi)+\Delta y\sin(\phi)]\\&\quad+K_3\cos(\theta)+K_4\sin(\theta).\end{aligned}\tag{57}$$

The trigonometric identity forms $\sin(\theta)=\frac{2\tan\left(\frac{\theta}{2}\right)}{1+\tan^2\left(\frac{\theta}{2}\right)}$ and $\cos(\theta)=\frac{1-\tan^2\left(\frac{\theta}{2}\right)}{1+\tan^2\left(\frac{\theta}{2}\right)}$ are substituted into (57) to produce (58),

$$\begin{aligned}&\cos(\phi)\left[\frac{1-\tan^2\left(\frac{\theta}{2}\right)}{1+\tan^2\left(\frac{\theta}{2}\right)}\right]+\sin(\phi)\left[\frac{2\tan\left(\frac{\theta}{2}\right)}{1+\tan^2\left(\frac{\theta}{2}\right)}\right]=K_1+K_2[\Delta x\cos(\phi)+\Delta y\sin(\phi)]\\&+K_3\left[\frac{1-\tan^2\left(\frac{\theta}{2}\right)}{1+\tan^2\left(\frac{\theta}{2}\right)}\right]+K_4\left[\frac{2\tan\left(\frac{\theta}{2}\right)}{1+\tan^2\left(\frac{\theta}{2}\right)}\right]\end{aligned}\tag{58}$$

and by multiplying both sides of the equality (57) by common denominator $1+\tan^2\left(\frac{\theta}{2}\right)$ and algebraically simplifying:

$$\begin{aligned}&\cos(\phi)-\cos(\phi)\tan^2\left(\frac{\theta}{2}\right)+2\sin(\phi)\tan\left(\frac{\theta}{2}\right)=K_1+K_1\tan^2\left(\frac{\theta}{2}\right)\\&+K_2[\Delta x\cos(\phi)+\Delta y\sin(\phi)]+K_2[\Delta x\cos(\phi)+\Delta y\sin(\phi)]\tan^2\left(\frac{\theta}{2}\right)\\&+K_3-K_3\tan^2\left(\frac{\theta}{2}\right)+2K_4\tan\left(\frac{\theta}{2}\right).\end{aligned}\tag{59}$$

Hereafter, to find a root for θ as independent variable, common terms are factorized and Eq. (59) is equated to zero to obtain

$$\begin{aligned}&\tan^2\left(\frac{\theta}{2}\right)[K_1 + K_2[\Delta x \cos(\phi) + \Delta y \sin(\phi)] - K_3 + \cos(\phi)]\\&+\tan\left(\frac{\theta}{2}\right)[2K_4 - 2\sin(\phi)] + K_1 + K_2[\Delta x \cos(\phi) + \Delta y \sin(\phi)]\\&+K_3 - \cos(\phi) = 0,\end{aligned} \tag{60}$$

in order to reduce expression complexity, the following Definitions are stated:

Definition 4.1 [term P]. Let P be defined as a Cartesian horizontal decreasing segment by expression:

$$P = K_1 + K_2[\Delta x \cos(\phi) + \Delta y \sin(\phi)] - K_3 + \cos(\phi). \tag{61}$$

Definition 4.2 [term Q]. Let Q be defined as a Cartesian vertical segment that is expressed by:

$$Q = 2K_4 - 2\sin(\phi) \tag{62}$$

and

Definition 4.3 [term P]. Let R be defined as a Cartesian horizontal increasing segment by expression:

$$R = K_1 + K_2[\Delta x \cos(\phi) + \Delta y \sin(\phi)] + K_3 - \cos(\phi) \tag{63}$$

Therefore, by substituting the terms of Definitions 4.1–4.3 into Eq. (60), the following quadratic form is deduced,

$$P\tan^2\left(\frac{\theta}{2}\right) + Q\tan\left(\frac{\theta}{2}\right) + R = 0. \tag{64}$$

By analytically solving (64), which is a second-degree equation using the general form, a real solution for θ is possible, thus

$$\theta = 2\arctan\left(\frac{-Q \pm \sqrt{Q^2 - 4PR}}{2P}\right) \tag{65}$$

Author details

Santiago de J. Favela Ortíz[†] and Edgar A. Martínez García*[†]
Laboratorio de Robótica, Instituto de Ingeniería y Tecnología, Universidad Autónoma de Ciudad Juárez, Chihuahua, Mexico

*Address all correspondence to: edmartin@uacj.mx

† These authors contributed equally.

References

[1] Azzabi A, Nouri K. Path planning for autonomous mobile robot using the potential field method. In: Proceedings of International Conference on Advanced Systems and Electric Technologies (IC_ASET 2017). 2017. pp. 389-394

[2] Reyes Muñoz JU, Martínez-García EA, Rodriguez Jorge R, Torres-Córdoba R. WMR kinematic control using underactuated mechanisms for goal-direction and evasion. In: Gorrostieta E, editor. Kinematics. IntechOpen; 2017. pp. 147-169. DOI: 10.5772/intechopen.70811

[3] Castro-Jiménez LE, Martínez-García EA. Thermal image sensing model for robotic planning and search. Sensors. 2016;**16**(8):1253

[4] Norouzi M, Valls Miro J, Dissanayake G. Probabilistic stable motion planning with stability uncertainty for articulated vehicles on challenging terrains. Autonomous Robots. 2016;**40**(2):361-381

[5] Kamalova A, Kim KD, Lee SG. Waypoint mobile robot exploration based on biologically inspired algorithms. IEEE Access. 2020;**8**: 190342-190355

[6] Ma L, Xue J, Kawabata K, Zhu J, Ma C, Zheng N. Efficient sampling-based motion planning for on-road autonomous driving. IEEE Transactions on Intelligent Transportation Systems. 2015;**16**(4):1961-1976

[7] Bai J, Lian S, Liu Z, Wang K, Liu D. Deep learning based robot for automatically picking up garbage on the grass. IEEE Transactions on Consumer Electronics. 2018;**64**(3):382-389. DOI: 10.1109/TCE.2018.2859629

[8] Fang B, Ding J, Wang Z. Autonomous robotic exploration based on frontier point optimization and multistep path planning. IEEE Access. 2019;**7**:46104-46113

[9] Lu C, Shan L, Jiang C, Dai Y. Reinforcement based mobile robot path planning with improved dynamic window approach in unknown environment. Autonomous Robots. 2021;**45**(1):51-76

[10] Zakiev A, Lavrenov R, Magid E, Svinin M, Matsuno F. Partially unknown environment exploration algorithm for a mobile robot. Journal of Advanced Research in Dynamical and Control Systems. 2019;**11**(8):1743-1753

[11] Yoon HS, Park TH. Motion planning of autonomous mobile robots by iterative dynamic programming. Intelligent Service Robotics. 2015;**8**(3): 165-174

[12] Qureshi AH, Ayaz Y. Potential functions based sampling heuristic for optimal path planning. Autonomous Robots. 2016;**40**(6):1079-1093

[13] Receveur JB, Victor S, Melchior P. Autonomous car decision making and trajectory tracking based on genetic algorithms and fractional potential fields. Intelligent Service Robotics. 2020;**13**(2):315-330

[14] Ferri G, Manzi A, Salvini P, Mazzolai B, Laschi C, Dario P. DustCart, an autonomous robot for door-to-door garbage collection: From DustBot project to the experimentation in the small town of Peccioli. In: 2011 IEEE International Conference on Robotics and Automation; 9–13 May 2011; Shanghai, China: IEEE; 2011. pp. 655-660. DOI:10.1109/ICRA.2011.5980254

[15] Han SI, Lee JM. Balancing and velocity control of a unicycle robot based on the dynamic model. IEEE Transactions on Industrial Electronics. 2015;**62**(1):405-413

[16] Klemm V, Morra A, Gulich L, Mannhart D, Rohr D, Kamel M, et al. LQR-assisted whole-body control of a wheeled bipedal robot with kinematic loops. IEEE Robotics and Automation Letters. 2020;**5**(2):3745-3752

[17] Li CG, Zhou L, Chao Y. Self-balancing two-wheeled robot featuring intelligent end-to-end deep visual-steering. IEEE/ASME Transactions on Mechatronics. 2021; **26**(5):2263-2273

[18] Jiang L, Qiu H, Wu Z, He J. Active disturbance rejection control based on adaptive differential evolution for two-wheeled self-balancing robot. Proceedings of the 28th Chinese Control Decision Conference (CCDC 2016). IEEE; 2016. pp. 6761-6766

[19] Chen X, Zhao H, Sun H, Zhen S. Adaptive robust control based on Moore-Penrose generalized inverse for underactuated mechanical systems. IEEE Access. 2021;**7**:157136-157144

[20] Iwendi C, Alquarni MA, Anajemba JH, Alfakeeh AS, Zhang Z, Bashir AK. Robust navigational control of a two-wheeled self-balancing robot in a sensed environment. IEEE Access. 2019;**7**:82337-82348

[21] Zhang C, Liu T, Song S, Meng MQH. System design and balance control of a bipedal leg-wheeled robot. International Conference on Robotics and Biomimetics (ROBIO 2019). IEEE; 2019. pp. 1869-1874

[22] Kim S, Kwon SJ. Dynamic modeling of a two-wheeled inverted pendulum balancing mobile robot. International Journal of Control, Automation and Systems. 2015;**13**(4):926-933

[23] Yue M, An C, Sun JZ. An efficient model predictive control for trajectory tracking of wheeled inverted pendulum vehicles with various physical constraints. International Journal of Control, Automation and Systems. 2018; **16**(1):265-274

[24] Tayefi M, Gen Z. Self-balancing controlled Lagrangian and geometric control of unmanned mobile robots. Journal of Intelligent and Robotic Systems: Theory and Applications. 2018; **90**(1–2):253-265

[25] Esmaeili N, Alfi A, Khosravi H. Balancing and trajectory tracking of two-wheeled mobile robot using backstepping sliding mode control: Design and experiments. Journal of Intelligent and Robotic Systems: Theory and Applications. 2017;**87**(3–4):601-613

[26] Sato R, Miyamoto I, Sato K, Ming A, Shimojo M. Development of Robot legs inspired by bi-articular muscle-tendon complex of cats. International Conference on Intelligent Robots and Systems (IROS). IEEE; 2015